US ARMY TECHNICAL MANUAL

PRINCIPLES OF COLD WEATHER CLOTHING AND EQUIPMENT

1944 WORLD WAR II
CIVILIAN REFERENCE EDITION

THE UNABRIDGED WARTIME HANDBOOK ON ADVANCED WINTER TECHNIQUES AND PROTECTION AGAINST THE ELEMENTS

U.S. WAR DEPARTMENT

Doublebit Press

New content, introduction, cover design, and annotations:
Copyright © 2021 by Doublebit Press. All rights reserved.
www.doublebitpress.com | Cherry, IL, USA

Original content under the public domain; unrestricted for civilian distribution. Originally published in 1944 by the U.S. War Department.

A Part of the Military Outdoors Skills Series

Doublebit Press Civilian Reference Edition ISBNs
Hardcover: 978-1-64389-180-4
Paperback: 978-1-64389-181-1

WARNING: Some of the material in this book may be outdated by modern safety standards. This antique text may contain outdated and unsafe recreational activities, projects, or mechanical, electrical, chemical, or medical practices. Any use of this book for purposes other than historic study may result in unsafe and hazardous conditions and individuals act at their own risk and are responsible for their own safety. Doublebit Press, its authors, or its agents assume no liability for any injury, harm, or damages to persons or property arising either directly or indirectly from any content contained in this text or the activities performed by readers. Remember to be safe with any activity or work you do and use good judgement by following proper health and safety protocols. In addition, because this book was from a past time and is presented in an unabridged form, the contents may be culturally or racially insensitive. Such content does not represent the opinions or positions of the publisher and are presented for historical posterity and accuracy to the original text.

DISCLAIMER: Doublebit Press has not tested or analyzed the methods, materials, and practices appearing in this public domain text and provides no warranty to the accuracy and reliability of the content. This text is provided only as a reprinted facsimile from the unedited public domain original as first published and authored. This text is published for historical study and personal literary enrichment purposes only and should only be used for such. The publisher assumes no liability for any injury, harm, or damages to persons or property arising either directly or indirectly from any information contained in this public domain book or activities performed by readers.

The Military Outdoors Skills Series
Historic Field Manuals and Military Guides on Outdoors Skills and Travel

Military manuals contain essential knowledge about outdoors life, thriving while in the field, and self-sufficiency. Unfortunately, many great military books, field manuals, and technical guides over the years have become less available and harder to find. These have either been rescinded by the armed forces or are otherwise out of print due to their age. This does not mean that these manuals are worthless or "out of date" – in fact, the opposite is true! It is true that the US Military frequently updates its manuals as its protocols frequently change based on the current times and combat situations that our armed services face. However, the knowledge about the outdoors over the entire history of military publications is timeless!

By publishing the **Military Outdoors Skills Series**, it is our goal at Doublebit Press to do what we can to preserve and share valuable military works that hold timeless knowledge about outdoors life, navigation, and survival. These books include official unrestricted texts such as army field manuals (the FM series), technical manuals (the TM series), and other military books from the Air Force, Navy, and texts from before 1900. Through remastered reprint editions of military handbooks and field manuals, outdoors enthusiasts, bushcrafters, hunters, scouts, campers, survivalists, nature lore experts, and military historians can preserve the time-tested skills and institutional knowledge that was learned through hard lessons and training by the U.S. Military and our expert soldiers.

Soldiers were the original campers and survivalists! Because of this, military field manuals about outdoors life contain essential knowledge about thriving in the wilds. This book is not just for soldiers!

This book is an important contribution to outdoors literature and has important historical and collector value toward preserving the American outdoors tradition. The knowledge it holds is an invaluable reference for practicing skills related to thriving in the outdoors. Its chapters thoroughly discuss some of the essential building blocks of outdoors knowledge that are fundamental but may have been forgotten as equipment gets fancier and technology gets smarter. In short, this book was chosen for Historic Edition printing because much of the basic skills and knowledge it contains could be forgotten or put to the wayside in trade for more modern conveniences and methods.

Although the editors at Doublebit Press are thrilled to have comfortable experiences in the woods and love our high-tech and light-weight equipment, we are also realizing that the basic skills taught by the old experts are more essential than ever as our culture becomes more and more hooked on digital technology. We don't want to risk forgetting the important steps, skills, or building blocks involved with thriving in the outdoors. This Civilian Reference Edition reprint represents a collection of military handbooks and field manuals that are essential contributions to the American outdoors tradition despite originating with the military. In the most basic sense, these books are the collection of experiences by the great experts of outdoors life: our countless expert soldiers who learned to thrive in the backwoods, deserts, extreme cold environments, and jungles of the world.

With technology playing a major role in everyday life, sometimes we need to take a step back in time to find those basic building blocks used for gaining mastery – the things that we have luckily not completely lost and has been recorded in books over the last two centuries. These skills aren't forgotten, they've just been shelved. *It's time to unshelve them once again and reclaim the lost knowledge of self-sufficiency.*

Based on this commitment to preserving our outdoors heritage, we have taken great pride in publishing this book as a complete original work. We hope it is worthy of both study and collection by outdoors folk in the modern era of outdoors and traditional skills life.

Unlike many other photocopy reproductions of classic books that are common on the market, this Historic Edition does not simply place poor photography of old texts on our pages and use error-prone optical scanning or computer-generated text. We want our work to speak for itself, and reflect the quality demanded by our customers who spend their hard-earned money. With this in mind, each Historic Edition book that has been chosen for publication is carefully remastered from original print books, *with the Doublebit Civilian Reference Edition printed and laid out in the exact way that it was presented at its original publication.* We provide a beautiful, memorable experience that is as true to the original text as best as possible, but with the aid of modern technology to make as beautiful a reading experience as possible for books that are typically over a century old. Military historians and outdoors enthusiasts alike are sure to appreciate the care to preserve this work!

Because of its age and because it is presented in its original form, the book may contain misspellings, inking errors, and other print blemishes that were common for the age. However, these are exactly the things that we feel give the book its character, which we preserved in this Historic Edition. During digitization, we ensured that each illustration in the text was clean and sharp with the least amount of loss from being copied and digitized as possible. Full-page plate illustrations are presented as they were found, often including the extra blank page that was often behind a plate. For the covers, we use the original cover design to give the book its original feel. We are sure you'll appreciate the fine touches and attention to detail that your Historic Edition has to offer.

For outdoors and military history enthusiasts who demand the best from their equipment, the Doublebit Press Civilian Reference Edition reprint of this military manual was made with you in mind. Both important and minor details have equally both been accounted for by our publishing staff, down to the cover, font, layout, and images. It is the goal of Doublebit Civilian Reference Edition series to preserve outdoors heritage, but also be cherished as collectible pieces, worthy of collection in any outdoorsperson's library and that can be passed to future generations.

WAR DEPARTMENT TECHNICAL MANUAL
TM 10-275

PRINCIPLES OF COLD WEATHER CLOTHING AND EQUIPMENT

WAR DEPARTMENT • 26 OCTOBER 1944

United States Government Printing Office
Washington: 1944

WAR DEPARTMENT,
WASHINGTON 25, D. C., 26 October 1944.

[A.G. 300.7 (1 Jul 44).]

TM 10-275, Principles of Cold Weather Clothing and Equipment, is published for the information and guidance of all concerned.

BY ORDER OF THE SECRETARY OF WAR:

G. C. MARSHALL,
Chief of Staff.

OFFICIAL:

J. A. ULIO,
Major General,
The Adjutant General.

DISTRIBUTION:

As prescribed in paragraph 9a, FM 21–6; Bn & H (1); C in Arctic Z (10); C in Extreme Cold Temperate Z (10); C in Temperate Z 1 & 2 (2). (See map on page iii as guide for distribution to areas in Arctic and Temperate Zones.)

For explanation of symbols, see FM 21–6.

CONTENTS

	Paragraphs	Page
CHAPTER 1. CLOTHING.		
Section I. General	1–3	1
II. Basic principles for keeping warm	4–9	1
III. Body clothing	10–26	4
IV. Footwear	27–37	18
V. Handwear	38–43	28
VI. Headgear	44–54	31
CHAPTER 2. EQUIPMENT.		
Section I. Packing equipment	55–57	35
II. Sleeping equipment	58–59	43
III. Cooking equipment and food bags	60–65	48
IV. Making camp	66–69	55
V. Mountain climbing equipment	70–73	58
VI. Skiing equipment	74–83	64
VII. Snowshoes	84–88	75
VIII. Miscellaneous equipment	89–93	78

Dark shaded area—Temperate Z 1..........................C(2)
Medium shaded area—Temperate Z 2......................C(2)
Light shaded area—Temperate Z 3.........................No Distr.
Unshaded area in Arctic Z...................................C(10)
Unshaded area in Extreme Cold—Temperate Z..............C(10)
Unshaded area in Tropic Z..................................No. Distr.

CHAPTER 1

CLOTHING

Section I. GENERAL

1. PURPOSE. The purpose of this manual is to insure proper use, care, and maintenance of cold-weather clothing and equipment. More casualties in extremely cold climates and mountainous terrain have been caused by inadequate clothing and equipment and the improper use of adequate clothing and equipment, than have been caused by enemy forces. Each soldier must learn a special technique to enable him to live comfortably and safely in the field under the most adverse conditions. Ignorance and carelessness in using and caring for specialized clothing and equipment have far more serious consequences in the mountains and cold climates than elsewhere. If a soldier in Georgia uses the wrong kind of socks, he has to nurse a painful pair of blistered feet; if he makes a mistake about the kind of footwear to use in the winter in Alaska, or if he uses the right type incorrectly, his feet freeze and he is likely to lose several toes. Remember at all times that a soldier who has fallen a casualty to the weather is just as useless as one knocked out by enemy fire.

2. SCOPE. This manual deals with all important items of clothing and equipment used in cold climates and in the mountains. Not everything in the manual is issued to each man or to each unit, but all specialized items which are issued to an individual or to a unit should be checked against the appropriate section in this manual. (*The items marked with an asterisk (*) are limited standard, and will be issued only until exhausted.*)

3. REFERENCE. For information concerning cold weather operations as a whole, see FM 31–15 and TM 1–240.

Section II. BASIC PRINCIPLES FOR KEEPING WARM

4. PRINCIPLE OF INSULATION. Substances which offer little resistance to the transference of heat are known as good *conductors;* those which resist it are said to be good *insulators.* Suitable cold weather clothing is not warm in itself; it is merely a good insulator and a poor conductor of heat. The heat of the body is held in by the clothing, and prevented from escaping into the atmosphere. Since still air is an excellent insulater, the best cold-weather clothes are those which entrap a considerable amount of air. The warmth of a woolen sweater lies mainly in the thousands of tiny air cells between the woolen fibers; fur is warm because of the air trapped among the hairs. (Some furs, such as caribou, have hollow hair, which makes them even warmer.) Several thin layers of cloth are better than one thick, heavy, matted piece of material, for air pockets can form between the layers.

5. IMPORTANCE OF LOOSE APPAREL. *Cold-weather clothing must be roomy.* If it is tight, much of the insulating air will be squeezed out. Moreover, the pressure on the body will restrict the circulation of the blood, which must move freely if frostbite is to be avoided. Loose-fitting clothes are perhaps not as smart in appearance as most military clothing, but they are essential if casualties from the cold are to be avoided.

6. CLEANLINESS. a. Dirt and grease mat clothing and fill the air pockets which give it its warmth. Grease makes clothing a good conductor of heat and a poor insulator. Therefore, *dirty clothes are cold clothes.* Clothes which are to be worn in cold weather must therefore be kept as clean as possible.

b. Wash and rinse woolens in lukewarm water; if washed in hot water they will shrink badly.

7. IMPORTANCE OF AVOIDING UNNECESSARY PERSPIRATION. a. Moisture is a good conductor of heat. When clothing gets damp, the spaces previously occupied by that excellent insulator, still air, become filled with heat-conducting moisture, which allows body-heat to escape. Clothing must be kept dry if it is to furnish good insulation. Since perspiration is the one factor most likely to cause wet clothing, shed clothes before getting really hot. In cold weather, it is better to be slightly chilly and dry than to run the risk of sweating. *To stay warm, avoid getting hot.*

b. There is a second reason for avoiding perspiration. Evaporation of perspiration causes a great heat loss. Regardless of care taken, short periods of strenuous activity will always result in some perspiration. In such cases, loose clothing can be opened at the neck, sleeves, waist, and ankles to allow air to circulate directly on the skin and evaporate the perspiration before the body becomes overheated.

8. ESSENTIAL CHARACTERISTICS OF COLD-WEATHER CLOTHING.
a. All clothing should be—
   (1) Loose.
   (2) Clean.
   (3) Dry.
b. Outer clothing should be—

(1) *Windproof*, to prevent displacement of the insulating air held in air pockets in the clothing.

(2) *Water-repellent*, to keep light rain and snow from penetrating to the inner clothing. Do not wear waterproof clothing in cold weather, as it will prevent perspiration from evaporating and cause inner clothing to become wet.

c. Inner clothing should consist of several layers of some good insulating material, such as wool or pile.

9. DRYING CLOTHING IN COLD WEATHER. Even in the coldest weather it is impossible to avoid perspiring entirely. Keeping perspiration down to a minimum, and drying the clothing after it gets damp, are two of the biggest problems for the soldier operating in the cold. He must be very careful to make use of every opportunity to dry his clothing.

a. **Removal of frost.** Regardless of the care taken to prevent perspiration in cold weather, some moisture will always condense in clothing and boots in the form of frost. Most of this frost (usually deposited just inside the outer layer) may be brushed or beaten off before it has a chance to melt. Before entering quarters, take off the outer clothing, turn it inside out, and remove as much frost as possible. If boots are flexible, turn them inside out also, and brush and scrape them. Solid boots should have as much frost scraped out as possible. Never take outer clothing indoors unless there will be time enough to dry it thoroughly. Otherwise, the frost will melt, soak the clothing, and then form as ice when it is exposed again to cold air; and icy clothing is much colder than clothing with a layer of frost on it. Leave frosted clothing outside if planning to return soon into the cold.

b. **Drying clothing indoors.** Hang each piece of clothing separately, where dry circulating air will strike it. Never put articles close to a fire or hot pipes. Leather articles, especially boots, must be dried slowly, as too much heat will ruin them. When the stove is lighted, hang clothes near the top of the tent, unless there is something steaming on the stove. Be careful not to dry clothes too close to a fire, as there is always danger of burning them; and make sure that if they fall, they will not land in the fire. A considerable amount of clothing has been destroyed in the past in these ways. Soldiers have even burned the boots and clothes they were wearing, while warming themselves by the fire. A soldier cannot get his feet warm before a fire when wearing several pairs of socks with cold-weather footgear. He will merely get them wet with melting frost, and be likely to burn his boots. Wet garments should be removed and thoroughly dried. At night, take damp mittens, socks, and other small articles into the sleeping bag to finish drying them by the heat of the body (see fig. 1).

c. **Drying clothing outdoors.** In good weather, it is not unusual to see water vapor rising from clothing, regardless of the temperature. To dry clothes, expose them to sun and wind, unless the snow is drifting, by hanging them from the guy ropes of tents or from the branches of trees.

Figure 1. Methods of drying clothing.

d. **Drying clothes on the march.** Mittens, socks, etc., may be hung from the pack, provided they are firmly enough attached not to fall off and get lost. Such articles may also be placed underneath the outer clothing if there is adequate ventilation to carry off the moisture as they dry.

Section III. BODY CLOTHING

10. **GENERAL CONSIDERATIONS.** In cold climates, several layers of light or medium weight woolen clothing are usually worn at the same time, covered by a windproof outer shell. This permits air to be trapped between the different layers, and also enables the wearer to regulate his temperature easily by adding or removing single pieces of clothing. A belt or drawstring around the waist helps to regulate ventilation. No gaps should remain at the neck, wrists, waist, or ankles. In general, the amount of clothing worn should depend on the temperature and on individual requirements.

11. JACKET, FIELD, M1943. a. Description. The field jacket is a windproof, water-repellent jacket which is used in temperate and cold climates as the necessary windproof outer shell. It may be worn over a shirt, sweater, and pile jacket, or any combination of the three. There are two breast pockets and two lower pockets which have unusual capacity. Although it is not intended that these be filled at all times, they are extremely useful in emergencies and at times when the pack is not carried.

b. Fitting. The field jacket should be large enough so that when the pile jacket or sweater is worn under it, the jacket will still be loose, and will not constrict the wearer in any way.

c. Use. Although the jacket must be loose enough to fit over the pile jacket or sweater, it can still fit properly when worn over a shirt alone.

Figure 2. Three possible collar adjustments of jacket, field, M1943.

This adjustability in size is made possible by the double sets of buttons at the wrists and neck, and by the waist drawstrings. There is another, more important, reason for these adjustable closures: they permit the wearer to adjust the warmth of his clothing. He can thus keep adequately warm without becoming overheated. Before he gets overheated and begins to perspire, he should loosen the waist closure, collar and sleeves, and let in cold air. If he gets cold, he should close the jacket at these points. If this is insufficient, he must vary the amount of clothing worn underneath at the first opportunity.

(1) *Collar.* The collar may be worn in three different ways:

(*a*) It may be worn open with the top button undone and with lapels turned back (see fig. 2).

Figure 3. Waist drawstring loop of jacket, field, M1943, may be easily untied to allow adjustment.

(*b*) It may be closed, with the top buttonhole on the farther button, if light clothing is being worn underneath; or on the nearer button, if the pile jacket, a muffler or other heavy clothing is being worn (see fig. 2).

(*c*) It may also be worn with the collar turned up and the flap buttoned under the chin (see fig. 2). When not in use, this flap must always be carefully buttoned out of the way. The position of the buttons should be checked and corrected if necessary.

(2) *Sleeve cuff.* There are two adjustment buttons on the cuffs. If the cuffs wear out before the rest of the jacket, they can be replaced.

(3) *Waist drawstring.* The drawstring usually is pulled until the

jacket fits tightly enough around the waist. Loops are then tied in the drawstring to keep it from pulling back through the eyelets. The extra material gathered up by the drawstring is distributed around the jacket. These loops must undo very simply so that it will not be difficult to

Figure 4. Methods of wearing clothing in layers. Pile jacket or pile parka is always covered by field jacket or cotton parka.

readjust the waist closure for colder or warmer conditions (see fig. 3). In rather warm weather the drawstring may be tied across the body and the jacket tied open to prevent flapping in the wind.

d. Washing precautions. The field jacket may be washed. It has been treated to make it water-repellent. Although the substance used is comparatively fast, the Army laundry must provide for replacing water-repellency whenever possible.

12. JACKET, FIELD, PILE. a. Use. The pile jacket (see fig. 4) is extremely warm when worn under the field jacket or parka.

b. Cleaning. The pile jacket may be washed in the same way as any woolen garment. Lukewarm water should be used. Never use hot water.

13. PARKA, FIELD, COTTON, OD. a. Nature and use. The olive-drab cotton field parka is well adapted to windy and cold climates. It is used as part of the windproof outer shell in climates which are too severe for the field jacket. The parka is a long-skirted jacket with a hood, with closures at the neck, wrists, and waist. When closed, these make it very warm; but they may be opened to permit adequate ventilation of the clothing when the weather is milder. There is a large pouch pocket on the front which has two openings. The olive-drab parka is often worn over the pile parka in extremely cold weather, and over other types of clothing in less severe weather.

b. Fit. The parka should fit loosely over the clothing worn beneath it. It must not constrict the wearer in any way, even when he has on the maximum amount of clothing (woolen undershirt, flannel shirt, high-necked sweater, pile jacket, pile parka). There are three sizes: small, medium, and large.

c. Manner of wear. When wearing a parka (or any other kind of cold-weather clothing), a soldier should take particular care to keep warm but avoid getting hot enough to perspire, since damp clothing leads to chilling. Before setting out in the morning, he should try to estimate what temperatures he may encounter and dress accordingly. During the day, however, he will have to make adjustments when temperatures change, or when periods of strenuous physical activity are followed by inactivity, or vice-versa. The wearer of a parka can easily regulate the warmth of his clothing in the following two ways:

(1) *Increasing or decreasing the amount of ventilation through clothing.* When the parka is closed at the neck, wrists, and waist very little air can circulate through the clothing. When it is left open at these places, circulating air hits much of the body surface, cooling it and evaporating the perspiration. After some experience, it is easy to regulate the amount of air needed to keep from perspiring. If sudden exertion makes the wearer too warm, he may loosen the waist drawstring and the neck of the parka. With the hood still up or thrown back on his shoulders, he can grasp the opening at the neck with both hands

and pump air into his clothing by raising and dropping his elbows (see fig. 5). A word on how to regulate each of these closures follows:

(*a*) *Neck closure.* For the greatest warmth in cold weather, the button which holds the flap in place must be fastened, as well as the three buttons along the edge of the opening. In warmer weather, these buttons may or may not be fastened as desired.

(*b*) *Sleeve closure.* The sleeves may hang loose or may be closed by means of the tabs, which give different adjustments, depending on whether they run around the front or back of the wrist.

(*c*) *Waist drawstring.* The wooden ball on the waist drawstring holds it as tight as desired. The skirt of the parka is always worn on the outside, over the trousers.

(*d*) *Drawstring at bottom edge.* The bottom drawstring is ordinarily used only by casualties and by inactive men who are forced to remain quiet for long periods in the cold. The drawstring is held in place by a wooden ball.

Figure 5. Cooling the body by allowing air to ventilate the clothing.

(2) *Removing or adding clothing.* If control over the ventilation of the clothing is insufficient, the wearer should change the amount of clothing worn at the first opportunity. Since it is extremely difficult to remove or add a pair of trousers when under way, he should alter clothing on the upper part of his body and then replace his parka. Although no set rules can be given for the correct quantity of clothing, experience and common sense soon teach the wearer how to stay warm enough and yet perspire very little.

d. Hood. The hood may be worn in several ways. If the snaps at the back of the hood are left loose, the hood will extend forward, creating a dead air space in front of the face (see fig. 6). If the weather is warmer or if better visibility is essential, the hooks may be snapped to

Figure 6. Methods of wearing the parka hood.

pull the hood back onto the head (see fig. 6). There are several ways to vary the shape of the hood: the upper set of snaps may be fastened, or the lower set, or both; or the upper snaps may be fastened onto the lower snaps. In any case, the drawstring will permit further adjustments, allowing ventilation around the neck or giving complete closure. In warmer weather, the hood may be thrown back off the head (see fig. 6), but in bad weather it must be tucked down the neck to avoid catching drifting snow or rain. Other headgear is often worn inside the parka hood, especially when the pile parka is not worn. The steel helmet may be worn over the hood, but the headband of the helmet liner must be adjusted before the need for wearing the helmet arises.

e. Warming the hands. In cold weather, the arms may be pulled out of the sleeves of the parka to warm the hands against the body or under the armpits (see fig. 7).

Figure 7. The arms may be drawn out of the sleeves into the parka to warm the hands against the body.

f. Washing. The parka may be washed. Although the substance used to make the parka water-repellent is comparatively fast, the Army laundry must provide for replacement of the water-repellent.

14. PARKA, FIELD, PILE. a. Nature and use. The pile field parka is used in regions of extreme cold. It is not windproof, and therefore is always worn, hair side in, underneath the olive-drab cotton field parka. The two parkas are worn as a unit, and are ordinarily put on and taken off together without removing one from the other. Three buttons will be found on the pile parka hood, which are buttoned into the corresponding buttonholes in the olive-drab parka hood. When the two hoods are attached, the snaps on the outer parka control both.

b. Care of fur ruff. The fur ruff around the face opening of the parka helps keep the face from getting frostbitten; it is particularly valuable in a crosswind. In cold weather, hoarfrost forms on the ends of the guardhairs, and should be stripped off with the fingers from time to time to prevent its accumulation. If the ruff becomes damp, it must be dried slowly to keep the furs from being ruined. The pile parka should be dry-cleaned, not washed.

c. **Removal of frost.** Regardless of the care taken to ventilate clothing, some perspiration will form and condense as frost between the layers. At the end of the day, or before entering a heated place for only a short time, the wearer should take off his parka and beat or brush off as much frost as possible. If he does not intend to stay in the warm place long enough to allow the remaining frost to melt and dry completely, he should leave this outer clothing outdoors. Otherwise the frost will melt, soak the clothes, and freeze as ice when he comes out again into the cold.

d. **Delousing.** Delousing is a comparatively simple matter in extremely cold weather. If infested garments are hung out for several hours, on 2 or 3 successive days, at temperatures approaching 0°F., all lice and eggs will freeze to death and may be shaken out of the clothing. Care must be taken, however, to insure that there are no folds which are not exposed to the frigid air.

15. **PARKA, FIELD, OVER, WHITE. a. Purpose.** This parka is used with white overtrousers and white overmittens to camouflage the soldier on snowy terrain. Unless it is kept absolutely clean, he will stand out against a white background. Therefore, it is used only for camouflage purposes, and *never* as a substitute for any other parka. Whenever the situation permits, a dirty overparka should be washed. When not in use, hang it in a clean, dry place.

b. **Use.** This garment is designed to fit over a cotton field parka. If a pile field parka is worn under the cotton field parka, the buttons on its hood are buttoned through the corresponding buttonholes of the cotton field parka, and then through those of the white overparka. The three parkas may then be worn as a unit, and can be put on and taken off together without removing one from the other. When the hoods are attached, the snaps on the outer parka control all three. If no pile parka has been issued, the soldier may find it advisable to sew buttons to the hood of the cotton field parka near the buttonholes, to be able to control both parkas with a single set of snaps. Otherwise, the inner parka should generally be left snapped back, while the outer is adjusted to meet weather conditions. (For further details on the use of the parka, see par. 13.)

16. **PARKA, WET WEATHER, AND TROUSERS, WET WEATHER. a. General.** When the temperature remains near freezing, and rain, sleet, and snow fall nearly every day, the soldier untrained to such conditions may find it harder to keep warm then in moderate or extreme cold. It is particularly important to keep dry in wet cold regions. When clothing becomes damp or wet, it is no longer efficient in keeping the soldier warm, and may chill his whole body. Such chilling of the body very often affects the feet, especially if socks and boots are also wet. Even at

temperatures above freezing, damp clothing and footgear have caused hundreds of men to become needless casualties, have necessitated amputations, and have even resulted in loss of life. There are two principal sources of moisture which are likely to wet clothing in such climates: precipitation and sweat. Rain, sleet, and snow must be guarded against to keep them from wetting the clothing from the outside. Perspiration must be held to a minimum to keep the clothing dry from the inside. Constant care is needed since it is very difficult to dry clothes in a wet cold region after they have become damp.

b. Use. When the soldier is not particularly active, he finds it rather easy to keep dry merely by wearing the wet-weather parka and trousers over adequate clothing. His feet will probably get best protection from shoepacs. He will find it more difficult to keep dry, however, when he is active. Unless proper care is taken, he may find that the waterproof wet-weather unit will trap so much perspiration that he will be wetter than he would have been if he had not worn the suit. Under such conditions, the soldier must learn from experience when it is best to remove the suit before starting to do active work. He must also learn under what conditions it is preferable to wear some water-repellent clothing, such as a field jacket or parka, in place of the waterproof wet-weather suit. Many men have successfully solved the problem by having two sets of clothing, one of which is always kept dry. The wet one is removed after coming into a shelter, and the dry one put on in its place. Even though every effort is made to dry it before going out again, it may still be damp by the time it is needed. In that case, it should be worn for the active outside work, so that the other set will still be dry when it is needed by the soldier at rest. In some instances, Hat, Rain,* Jacket, Rain,* and Trousers, Rain,* are issued in place of the wet-weather parka and trousers.

c. Precautions against perspiration. When the wet-weather parka and trousers are worn, it is essential to start with dry clothing underneath. As little underclothing as possible should be worn, so that sweating will be kept to a minimum. The moisture from perspiration will wet clothing less if the suit is well ventilated. The ventilating eyeholes, although helpful, will not provide sufficient ventilation. The drawstring in the parka hood and at the waist, the buckles at the waist of the trousers and at the neck, and the snaps at the wrists may be left undone to allow air to circulate through the clothing. The snaps at the ankles may be left loose when there is no danger of stepping into water, mud, or snow. This ventilation helps to reduce the body temperature by allowing the warm air around the body to escape, and by facilitating evaporation of perspiration.

d. Necessity for an extra change of clothing. In wet cold regions, it is very difficult to dry clothing anywhere, and almost impossible when away from a heated shelter. On an extended trip away from the base, each soldier should carry at least a complete change of socks, woolen underwear, mitten-inserts and insoles.

e. **Special notes on Parka, Wet-Weather.** This is not a substitute for water-repellent parkas such as the Parka, Field, Cotton, OD, or the Parka, Ski, Reversible,* which are intended for cold, windy weather. As its name suggests, the wet-weather parka is intended only for *wet* conditions.

f. **Special notes on Trousers, Wet-Weather.** These trousers have a high bib and are held up by attached suspenders. Normally, the waist buckles will be adjusted to fit the wearer, but if he gets too warm he may open them to allow greater ventilation, letting the trousers hang on the suspenders. There are two ankle adjustments. The trouser legs may be worn either inside or over the boots, depending on conditions. In very wet weather, it is often preferable to leave the trousers outside the boots, so that water will not run down the legs into the boots. For crossing shallow water, a more watertight closure may be made by tying the trouser legs tightly around the boottops. Strips of cloth several inches wide and 3 or 4 feet long may be wrapped around the tops of the boots or around the cuffs of the trousers, as puttees.

17. OVERCOAT, PARKA TYPE, WITH PILE LINER. a. Use. The parka type overcoat with pile liner is used in cold climates, particularly by individuals whose work is fairly inactive, such as drivers and sentries. For strenuous activity, other types of parkas are more practical. The pile liner may be removed and the outer layer worn alone, to fit variations in temperature.

b. **Types.** (1) *Overcoat, Parka Type, with Pile Liner.* This is a two-piece garment consisting of an outer shell with a pile-lined hood, made of windproof material, and an inner pile liner. The pile liner snaps into the outer shell (see fig. 4).

(2) *Overcoat, Parka Type, Reversible, With Pile Liner.* This type has two distinct parts, the windproof shell and the pile lining. It is possible to reverse the shell without removing the lining, exposing either the olive-drab or the white side for the best camouflage effect. Unless the white side is kept clean, the coat will stand out clearly against a snowy background. The olive-drab side is therefore worn on the outside when there is no need for camouflage.

18. JACKET, MOUNTAIN*. a. Nature and use. The mountain jacket, which is issued to mountain troops, serves either as a parka or as an ordinary jacket. It should have a loose fit, especially around the arms and shoulders.

b. **Parts.** (1) *Hood.* The hood is adjusted to give a good closure about the neck and face by fastening the buttons and adjusting the drawstring around the face. If the hood is not needed, it fits neatly into a pocket made for the purpose if folded three times and tucked in flat. The jacket-collar covers the opening of the pocket.

(2) *Interior suspenders.* These suspenders are used to support weights in the pockets. Buckles are provided to adjust the suspenders to the wearer and to the loads he carries in his pockets. When the jacket is slipped on, one set of straps stays in front of the arm, allowing the suspenders to run directly over the shoulders. In warm weather, these interior suspenders serve a second purpose. If no pack is being worn, the jacket may be hung from the shoulders on the suspenders which will resemble rucksack shoulderstraps.

(3) *Pockets.* The pockets of the mountain jacket are designed to carry a considerable load. The slide fastener which holds the back pocket shut is on the right side. Bulky articles, folded flat, may be carried in the back pocket; more compact objects are placed in the front pockets. The weights in the front and back pockets should be equalized as much as possible, to keep the jacket from being dragged down either in front or in back. If there is a large load in the pockets, the belt is tightened, to keep it from flopping. If very little is being carried in the back pocket or if the load consists only of heavy compact articles, it may be advisable to tighten the belt first to close off the bottom portion of the pocket and to keep the load higher on the shoulders.

c. **Cleaning precautions.** The mountain jacket is treated to make it water-repellent. Although the substance used is comparatively fast, the Army laundry must provide for replacing the water-repellency when the jacket is cleaned or washed.

19. **PARKA, REVERSIBLE, SKI*.** a. **Nature and use.** This garment is well adapted to mountain weather. It is white on one side and olive-drab on the other. Either side may be worn on the outside, depending on which offers the better camouflage. If there is no need for camouflage, the olive-drab should be worn on the outside, as it absorbs more warmth from the sun than the white side, and the white side is kept cleaner when worn on the inside.

b. **Hood.** See paragraph 13d.

c. **Cleaning procedures.** Because dirt spoils the camouflage effect of the white side, it is important to keep it clean. If it should become spotted, clean it with soap and water, making absolutely sure that all the soap is carefully washed out, if even a small amount of soap is left, the water-repellent quality will be ruined. Never use leaded gasoline to remove spots, as it will leave a yellow stain. Ordinarily, the parka should be dry-cleaned. Although the substance used to make the parka water-repellent is comparatively fast, the Army laundry must provide for replacing the water-repellency.

20. JACKET, COMBAT, WINTER*. a. Purpose and use. The winter combat jacket is worn by members of the armored force when the weather is cold enough to warrant it. Since it will frequently be worn over other heavy clothing, it should be fitted loosely enough to cover a flannel shirt and sweater.

b. Cleaning precautions. The lining of this cotton jacket is made of wool. If possible, therefore, it should be dry-cleaned when it gets dirty, in order to get both the lining and the cotton outer material clean. If it has to be washed, it should be treated as a *woolen* and washed with soap and lukewarm water.

21. JACKET, FIELD, OD, ARCTIC*. a. Nature and use. The Arctic field jacket is useful in moderately cold temperatures. In extreme cold or wind some type of parka should be worn. The jacket should be large enough to fit comfortably over a flannel shirt and sweater.

b. Cleaning. Whenever possible, this jacket should be dry-cleaned rather than washed. If dry cleaning facilities are not available, it should be washed with lukewarm water and soap like other woolens. Very hot water will shrink the wool lining.

22. TROUSERS, FIELD, COTTON, OD. a. Nature and fitting. In cold weather, olive-drab cotton field trousers provide a windproof, water-repellent outer shell for use over heavy trousers. In warmer weather, they are used alone over summer underwear. The waist is somewhat oversized, but there is an adjustment tab on the waistband above the hips to make it fit well under all conditions. When trousers are fitted over woolen underwear, the tabs should be buttoned to the rear buttons to give a large waist measurement. When they are worn alone, the tabs should be buttoned to the front buttons to reduce the size of the waist. Regardless of the climate in which they are issued, the trousers must always be fitted to permit the use of heavy trousers (such as Trousers, Field, Wool, OD), underneath. Trousers which are marked size 32, for example, fit properly under all conditions on a man who normally wears a size 32.

b. Parts. Several parts warrant special considerations:

(1) *Belt loops, suspender loops.* Field trousers may be used with either a belt or suspenders. If additional trousers are being worn, the suspenders may be attached to the inner pair of trousers (see fig. 4). The suspender loops will then hold up the field trousers. If heavy loads are carried in the pockets, suspenders may be useful.

(2) *Ankle closure.* There are flaps at the inside of the ankle which may be buttoned onto the button at the outside of the ankle to provide good closure. The tightness of closure varies, depending upon whether the tab is placed in front of or behind the ankle. The cuffs of the

trousers may be worn inside or outside the boots, but in wet snow or other wet conditions it is usually better to wear them with the flap buttoned tightly over the outside of the boot to prevent water from running down the trouser legs and seeping into the boots. When the trousers are worn loose at the bottom, air can circulate up the trouser leg. The flap should then be buttoned flat.

c. Washing precautions. Field trousers may be washed but, whenever possible, they should be treated afterward to restore water-repellency.

23. TROUSERS, FIELD, WOOL, OD. These are heavy woolen trousers, worn sometimes alone, but for the most part, especially in cold weather, under olive-drab cotton field trousers, which are windproof and add a great deal of warmth. The trousers may hang loosely at the ankles, or be closed by means of the button tab. Use either a belt or suspenders. When not needed, the protective flap at the fly may be folded back and buttoned to the suspender buttons.

24. TROUSERS, FIELD, OVER, WHITE. a. Nature and use. These trousers are intended primarily for camouflage. Since they stand out against a snowy background when dirty, they must be kept as clean as possible. Therefore, although they add some warmth because of their windproof qualities, use them only where camouflage is needed. If possible, they should be washed even when only slightly dirty. Store them in a clean, dry place.

b. Manner of wear. These trousers fit over other trousers and may be slipped on and off without removing the boots. They are usually worn outside the footwear, so that they may be removed quickly when necessary. Adequate ankle-closure is given by the drawstrings or snaps at the ankles.

25. TROUSERS, MOUNTAIN*. a. Nature and use. Mountain trousers, which are issued to mountain troops, are not particularly warm in themselves, but they form a windproof shell over woolen underclothes or other trousers.

b. Parts. Several parts warrant special consideration:

(1) *Belt loops and suspender buttons.* Mountain trousers are equipped with both of these devices. The suspender buttons may be used to hold up woolen underpants.

(2) *Cuffs.* The cuffs at the ankles fit over all the socks. The straps under the insteps keep the cuffs from pulling out of the boots.

(3) *Pockets.* These must be carefully buttoned to keep out snow.

26. TROUSERS, KERSEY-LINED*, AND TROUSERS, COMBAT, WINTER*.
a. Nature and Use. Kersey-lined trousers combine the windproof outer

shell and the warmth-giving woolen layer in one garment. They may be worn by themselves or, more commonly, over woolen underwear. *Winter combat trousers* are used by members of the armored force in cold weather. The latter trousers come up high on the chest and are held up by suspender straps. They are adjusted to the height of the wearer by buckles on the straps. There is a good ankle-closure provided by snaps at the bottoms of the trouser legs.

b. **Cleaning precautions.** Since the trousers are lined with wool, they should be dry-cleaned. If they must be washed, they should be treated as woolens and washed with soap and lukewarm water to keep the lining from shrinking.

Section IV. FOOTWEAR

27. **GENERAL CONSIDERATIONS.** It is difficult to keep the feet warm in cold climates. The weight of the body compresses much of the air out of the insulating material under the foot, and the movement of the foot prevents the air from remaining still. Furthermore, a certain amount of perspiration always forms. In cold weather this goes, as water vapor, toward the outer layer, and escapes to a greater or lesser extent, depending on the waterproofing qualities of the boots. Since most boots are at least slightly waterproof, a certain amount of moisture collects inside them in the form of frost. In the most extreme cold, the temperature of the boots themselves is so cold that the moisture condenses inside the footgear no matter how permeable the boots are. Much of this frost forms in the insoles and outer pair of socks. However, precautions may be taken to insure adequate warmth for the feet.

28. **SOCKS. a. Types.** There are several types of socks for use in cold weather. These are worn in various combinations, depending on the type of boots worn. The shoes should be fitted with the proper sock combination, and should always be worn with this combination.

(1) *Wool, Ski.* These are heavy woolen socks which come about halfway up to the knees. They are worn either next to the skin or as outer socks.

(2) *Wool, Heavy.* These are lighter than Socks, Wool, Ski. They are ordinarily used alone or as inner socks.

(3) *Wool, Cushion-Sole.* These are lightweight socks with a reinforced sole. They are usually worn alone or as inner socks.

(4) *Felt.* Loose-fitting felt socks are used in mukluks, and should keep the feet warm in the coldest weather. They are the correct size if they will fit over two pairs of wool ski socks without constricting the feet in any way. Felt socks should be dry-cleaned and not washed.

(5) *Arctic.** These are about the same weight as Socks, Wool, Ski, but reach up to the knees.

b. Wear. (1) Several pairs of heavy socks are warmer than a single extremely thick pair, because more air can be trapped between the layers.

(2) When two or more pairs of socks are worn at the same time, the outer pair should be larger so that the feet will not be cramped and circulation restricted. If the size numbers become obliterated, mark the socks with some easily identified mark. For example, one strand of colored yarn, thread, or string may indicate the smaller socks, while two indicate the larger. If different colored yarn, thread, or string is available, each color may indicate a different size.

(3) Do not cram boots full of socks. For example if boots fit well with two pairs of socks and insoles, a third pair will only make the boots tight and cold and cause the feet to become frostbitten.

(4) During a long halt or at the end of the day, change damp socks as soon as possible. In cold weather, damp socks are dangerous because of the possibility of frostbite. If they become really wet, blisters on the feet often result regardless of the temperature.

(5) Dry socks carefully. Brush, scrape, or beat off any visible frost. If the socks are then exposed to sun and air, some moisture will pass off, even into frigid air. When socks are placed inside well-ventilated clothing, the heat of the body will help to dry them. Be careful not to burn them when drying them in front of a fire. In the field, the soldier takes damp socks into his sleeping bag with him at night, where body-heat will help to dry them.

(6) If possible, put on dry insoles and socks every morning.

(7) Dirty socks are cold socks. Socks should be washed in lukewarm water; hot water will shrink them. After washing, when the soap has been carefully rinsed out, the socks should be squeezed out and pulled into shape again. They must be dried thoroughly before they are worn.

29. INSOLES. a. Nature and use. Insoles are essential to cold-weather footwear, for they provide added insulation in the soles, thus preventing the cold from coming up from below and absorbing moisture from the feet. Two pairs of insoles, cut to the proper shape, are issued with all new cold-weather footwear. The two pairs are used alternately, one pair being dried while the other is used. In mukluks, two pairs are used for greater interior insulation.

b. Care. When insoles are damp, they lose much of their insulating quality; they should be dried at the first opportunity. Sun and air will help dry them, even when the temperature is below freezing. They will also dry if kept next to the body in well-ventilated clothing. When the soldier gets into his sleeping bag at night, if the insoles are still damp he should put them in with him to finish drying. Insoles must be removed the minute boots are taken off. If left in the boots, they will not dry properly and will soon freeze to the boots and prevent both from drying. Alternating pairs of insoles each morning will keep them in good condition.

c. **Replacements.** Replacements come in blocks of felt which must be cut out to fit the size of the boots in which they will be worn. The old insoles which were issued with the boots can serve as a pattern for cutting the new ones.

30. BOOTS. a. General. There are several types of special boots issued to mountain troops and to troops operating in extreme cold. Each has its particular use and is especially suited to specific conditions. Heavily greased boots, such as high-top blucher boots, are not satisfactory for really cold weather. They are very cold because the grease in them freezes and does not permit the boots to "breathe." Further, when the grease freezes, the boots become stiff and circulation is impaired. Very little dubbin should be applied to boots during the fall and winter. In cold weather, boots should be laced loosely, as tightness reduces insulation and restricts blood circulation. If there is no danger that snow will come over the tops of the boots, they should be left loose on the legs so that moisture can escape upward.

b. **Mukluk.** (1) *Nature and use.* Mukluk boots are the warmest type of footwear issued and may be worn in temperatures of 20°F. or lower. Since they are not waterproof, they do not collect much frost. For the same reason, however, they cannot be used in weather warmer than about 20° F. or in wet snow. They are not suitable for use on skis, except when the skis are equipped with flexible bindings.

(2) *Fitting.* Only three sizes of mukluks are needed, since they fit very loosely. The following sizes should be issued:

(a) *Small*—for feet up to size 7½.

(b) *Medium*—for feet from size 8 to 10.

(c) *Large*—for feet size 11 and up.

(3) *Wear.* In the coldest weather, a pair of cushion-sole socks, two pairs of wool ski socks, a pair of felt socks, and two pairs of insoles fill the mukluks, and should provide adequate protection against the cold. (See fig. 8 for correct sock combination.) It is sometimes more comfortable to wear the second pair of insoles inside the outer socks. Great care should be taken to set the insoles squarely in the mukluks, and to prevent their sliding forward and bunching at the toe when the foot is inserted. If the insole tends to slip to one side at the heel when marching, make an immediate adjustment. Mukluks are usually laced in the following manner (see fig. 8):

(a) Fasten the tie string loosely at the top to hold the mukluk up under the knee.

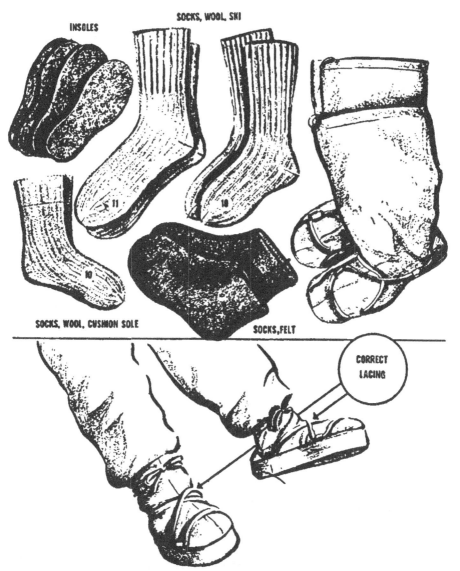

Figure 8. Correct sock combination for mukluk boots; 1 pair Socks, Wool Cushion Sole; 2 pairs Socks, Wool, Ski; 1 pair Socks, Felt; 2 pair insoles.

(*b*) Run the laces through the toe loops so that the ends run toward the heel, and are of equal length.

(*c*) Cross the laces over the instep and run them through the ankle and heel loops.

(*d*) Wrap the laces loosely two or three times around the ankle and tie in bowknots. Once the laces are placed in the loops, they may be loosened to remove the foot; they need not be pulled out.

(4) *Arch support.* When a change is made from stiff shoes which give good arch support, to flexible mukluks, a week may be required

to overcome muscular fatigue. The inconvenience sometimes experienced during the first week of wearing mukluks is caused by the sudden strenuous use of muscles which have not been strengthened sufficiently while wearing stiff leather shoes, and results in no injury. Soldiers should be given some time for their feet and leg muscles to get accustomed to the new footgear before undertaking long marches.

(5) *Care.* One of the reasons for the warmth of a mukluk boot is that it is not waterproof. Even if the temperature is well below freez-

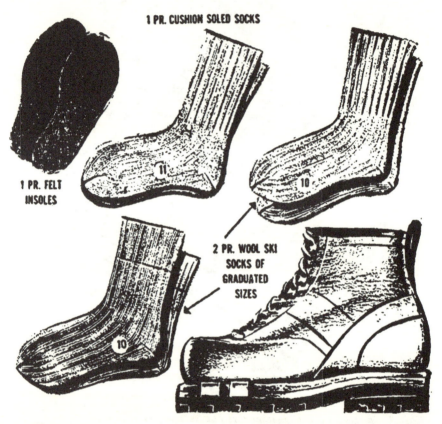

Figure 9. Correct sock combination for boots, ski-mountain; 1 pair socks, wool, cushion-sole; 2 pairs of socks, wool, ski, 1 pair insoles.

ing, perspiration will usually pass off into the air as water-vapor without forming frost in the boot or socks. For this reason, dubbin, which is water-repellent, should never be applied to a mukluk. If frost does form in a mukluk, the boot should be turned inside out, the frost brushed out, and the inside left exposed to the air. The moment mukluks are taken off, insoles should be removed to prevent them from freezing in. Snow should always be brushed off before mukluks are brought into a heated place.

31. **BOOTS, SKI-MOUNTAIN.** a. **Nature and use.** Ski-mountain boots (see fig. 9) combine the characteristics of hobnailed mountain-climbing boots and ski boots. Mountain troopers use them more than any other type of footwear, because they are suitable under almost any conditions. They are warm in all but the coldest temperatures. In the mountains, they are equally useful in winter and in summer. When nailed or fitted with rubber-cleated soles, they hold on both wet and dry rock, on steep grass, snow, and glacier ice. (On smooth ice, crampons should be worn.) Ski-mountain boots have square toes and the heels are grooved to fit ski bindings. The rubber-cleated soles make no noise and do not cause sparks, as nailed boots sometimes do. Rubber-cleated soles wear better than the nailed type; and boots made with this type of sole are warmer and lighter than those with nailed soles. The ski-mountain boot with nailed sole grips well on most surfaces. However, crampons should be worn on all steep, smooth ice. If hobnails or edge nails are used, missing ones should be replaced at the first opportunity. In the field it is usually necessary to place a rock inside the boot when hammering nails in. Boots can be nailed most easily when the sole is wet. Ski-mountain boots with plain leather soles may be used for skiing or walking on easy terrain but they are treacherous on grass or uneven ground.

b. **Wear.** Ski-mountain boots must fit snugly but not tightly over one pair of cushion-sole socks, two pairs of ski socks and a pair of felt insoles. (See fig. 9 for correct sock combination.) Although boots should be laced up very tightly when skiing downhill, for good ankle-support, the wearer must remember to loosen them again later for better circulation, especially if the weather is cold.

c. **Care.** Ski-mountain boots are perhaps the most important item of the mountain trooper's clothing. Without them he cannot hope to carry out his duties safely and efficiently. He must therefore take great care to keep them in good condition. The following measures should be taken as needed:

(1) When quarters are reached or camp is made, boots should be taken off and insoles removed to dry. *The boots should always be opened wide, and left, so that they can be slipped on easily even if frozen solid.* If the boots are used as a pillow or placed under the feet at night, they will freeze less easily. If they are solidly frozen in the morning, they should be slipped on before breakfast. By the time breakfast is over, they will be sufficiently thawed to lace up.

(2) Always dry boots slowly. They must never be left near a fire or near warm pipes, no matter how wet they are, because the heat will cause the leather to crack.

(3) Treat the boots with dubbin about once a week, if the weather is not too cold. In extreme cold, dubbin should not be used, since

grease freezes and makes boots colder. If the boots are used a great deal in wet snow, frequent applications are needed.

(4) Treat boots with dubbin in the following manner:

(*a*) Clean off mud and dirt.

(*b*) If the boots are wet, rub only enough moisture off the surface to allow the dubbin to spread out evenly. If the application is made while the boots are still damp, the dubbin will go on more evenly and the leather will not get stiff.

(*c*) Next, coat the whole boot (including the sole, if leather) with dubbin.

(*d*) Give special treatment to the welt, where the upper part of the boot and the sole meet.

(*e*) Rub the dubbin into the leather and polish the boots by hand.

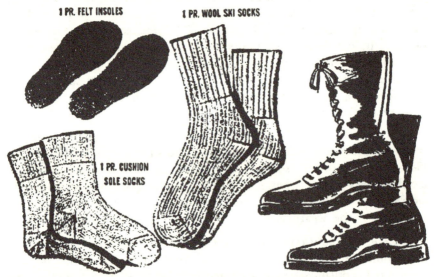

Figure 10. Correct sock combination for Boots, Blucher, High Top; one pair of socks, wool, cushion soles; one pair socks, wool, ski, one pair of insoles.

32. BOOTS, BLUCHER, HIGH TOP. a. Use. These boots (see fig. 10) are highly greased and are not intended for use in extreme cold. Although they are relatively waterproof, they do soak through when exposed to wet conditions for long periods of time. Once wet, they are very difficult to dry. These boots should therefore be used only in moderate weather, neither too wet nor too cold. Frequent applications of dubbin are needed to keep out as much moisture as possible. Blucher boots are designed to be worn with a pair of insoles, one pair of cushion-sole socks, and one pair of wool ski socks. (See fig. 10 for correct combination.)

b. Care. Dry blucher boots slowly, away from fire or hot pipes, which will ruin the leather. Treat them with dubbin at least once a week. For directions about applying dubbin see paragraph 31.

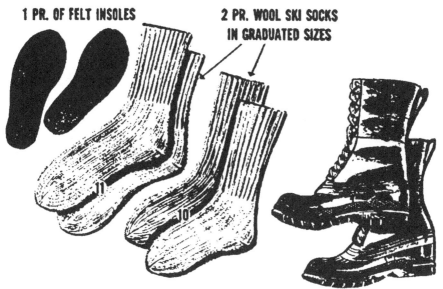

Figure 11. Correct sock combination shoepacs; two pairs socks, wool, ski, one pair insoles.

33. SHOEPACS. a. Nature and purpose. Shoepacs are boots with rubber feet and leather tops (see fig. 11). They are particularly useful in wet terrain or snow, and are excellent for use with snowshoes. They are waterproof boots and are not intended for protection against extreme cold. Since the feet are made of rubber, perspiration cannot escape easily and frost will form inside in cold weather, making the boots stiff and cold. They are adequate for active men in temperatures down to about 0° and for inactive men in temperatures down to about 15°, provided their socks and insoles are dry.

b. Fit. Shoepacs should fit snugly but not tightly over two pairs of wool ski socks and one pair of insoles (see fig. 11). Improper fit sometimes leads to chafing or blistering of the foot. (1) If the seam beneath the ankle bone chafes, try one of the following methods:

(a) Raise the foot slightly by putting a pad under the heel. The pad may be cut out of the piece of felt intended as an insole replacement.

(b) Turn the top of the outer sock down over the ankle to provide more padding.

(2) If the heel gets chafed, there is usually too much space through the instep. This can be corrected in one or more of the following ways:

(a) *Lace the shoepacs snugly over the instep and tie a square knot. Then lace the top loosely.* (See fig. 11.)

(b) Turn the top of the outer sock down to the instep, to take up extra space and pad the heel.

(3) Chafing is not always caused by improper fit. If the leather is not treated with dubbin it gets hard. If socks are dirty, they may chafe. If they are damp, the skin may become tender and blister easily.

c. Use. Socks and insoles used in shoepacs must always be kept as dry as possible. If they are damp, it is advisable to change them immediately at the beginning of a long halt, or at the end of the day, while the feet are still warm. The insoles are always removed when the shoepacs are not being worn. Shoepacs, like all boots, dry best when left right side up in a warm place.

d. Care. The leather tops can be kept in good condition by a weekly application of dubbin. When applying, care must be taken to keep the dubbin from getting onto the rubber and seam, or it will ruin the rubber. Perhaps once a month, when opportunity offers, the inside of the shoepacs should be washed out with warm water and soap, and then dried thoroughly.

e. Precautions before wearing leather boots. After wearing shoepacs for some time, it takes a short while to get reaccustomed to leather shoes. Whenever possible, leather shoes should be worn at first for only half a day at a time, and not on long marches.

34. OVERSHOES, ARCTIC. Arctic overshoes are not a substitute for the more waterproof type of footgear, such as shoepacs or rubber boots; but when worn over service shoes in wet or muddy terrain they will keep the feet fairly dry. They are not ideal marching boots, for they do not fit well enough over other shoes to allow all the footgear to be lifted as a single unit. In temperatures around freezing they give a certain amount of protection against the cold when worn over a service shoe. In really cold weather, cloth-topped overshoes should be worn over several pairs of woolen socks and insoles, or over felt shoes. If the heel cup is filled with part of an insole cut to the proper shape, the overshoe is more comfortable for such use. In very cold weather, do not wear service shoes in overshoes.

35. SHOES, ARCTIC, FELT. a. Purpose. Arctic felt shoes are used in extremely cold weather, and are worn without any outer covering (see fig. 12). They will keep the feet of active men warm at temperatures down to about minus 40°, and those of inactive men warm at temperatures down to about minus 20°; when the temperature is colder than this, mukluks are needed. Felt shoes give the foot more arch support than mukluks and are preferred by some men as cold-weather marching boots. They may be used with flexible cross-country bindings for skiing. Arctic felt shoes should not be worn when the temperature is above 20°, or the heat from the foot will melt the snow and wet the shoes. Always brush off any snow before going into a heated place.

b. Fit. Arctic felt shoes must fit snugly, but not tightly, with two pairs of wool ski socks and one of insoles (see fig. 12). Sometimes a cushion-sole or a heavy wool sock is worn next to the foot in addition to the other socks.

c. Drying. Felt shoes must be kept dry to keep the feet warm. Snow will not wet them until the temperature rises above 20°. They will, however, get somewhat damp from perspiration collecting in them as frost, and must therefore be dried carefully every night. Remove the insoles as soon as the shoes are taken off. Scrape off any frost which is visible. Then open shoes wide and leave them to dry. If you are camping out, put them where they will get the greatest amount of heat, such as near the peak of the tent. If indoors, keep them in a moderately warm place but away from fires, stoves, hot pipes, etc.

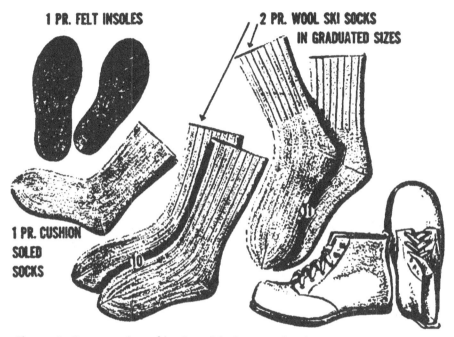

Figure 12. Correct sock combination with shoes, Arctic felt; one pair socks, cushion-sole; two pairs socks, wool, ski; one pair insoles.

36. SHOES, FELT, COLD CLIMATE, AND OVERSHOE, ARCTIC. a. Purpose and use. Shoes, Felt, Cold Climate, are used especially by garrison troops whose duties require them to come in and out of heated buildings, and by vehicle drivers, mechanics, etc. They are less suited to marching than the Arctic felt shoe. Because of the permeable nature of the felt, they are not uncomfortable in moderately warm rooms. When used outdoors, they should always be covered by special felt shoe overshoes to keep them from getting wet in slushy patches around entrances or in the floors of vehicles. These overshoes are specially sized to fit over the garrison felt shoe and should always be issued with them, since ordinary arctic overshoes are not of correct size for the purpose. They are adequate for active men in temperatures down to about minus 20° and for inactive men in temperatures down to about 0°.

b. Fit. Garrison felt shoes must fit snugly, but not tightly, with one pair of wool ski socks, one pair of cushion-sole socks, and one pair of felt insoles.

c. Drying. Like any other cold-weather footgear, garrison felt shoes must be dry to keep the soldier's feet warm. If they are worn indoors, the wearer must check carefully to see that his socks, insoles, and shoes are dry, before going outdoors for any considerable length of time. The shoes should be dried carefully every night after the insoles have been removed.

37. GAITER, SKI. a. Nature and use. Ski gaiters make a good closure between the trousers and the ski-mountain boots and prevent snow and dirt from working into the tops of the boots. The ski gaiter should be put on in the following manner:

(1) Place the gaiter on the foot with the lacings on the outer side.

(2) Run the longer of the two leather straps under the arch, over the top of the instep, and through the leather tunnel; then buckle firmly to the other strap.

(3) Start the lace at the rear bottom eye, tying a knot on its end to keep it from sliding through the hole.

(4) Slip the lace out through the outer eye on the bottom of the gaiter, causing the front half to be pulled over the back.

(5) Run the lace up the hooks to the top, where it is slipped in through the two superimposed eyes, and out through the rear top eye.

(6) Fasten the lace with a half-hitch, which will undo when pulled. A hard knot may become impossible to untie if ice freezes on it.

(7) Shove the loose end into the top of the gaiter.

b. Special precautions. The buckle which fastens the leather straps sometimes gets clogged with ice. If this is likely to happen, it is advisable to leave the end of the strap loose and not tuck it under the bar of the buckle. If it should get frozen, heat (a lighted match) may be applied, provided care is taken not to scorch the boot or gaiter.

Section V. HANDWEAR

38. COLD-WEATHER HANDWEAR. The same principles that apply to keeping warm in other types of clothing hold true for handwear.

a. Gloves must be clean. Dirt, especially grease, ruins insulation and helps conduct heat away from the hands.

b. Gloves must be dry. If handwear is damp, it loses its insulating qualities. Remove a layer before the hands begin to sweat. If bare hands get wet from melting snow, dry them carefully before putting them back into mittens. The amount of protection needed to keep the hands warm varies with the individual and his duties. Mittens are warmer than gloves, since the fingers are together and help to keep each other warm. Delicate work, however, requires fingered gloves or bare hands. Handle metal with care at cold temperatures for the skin may stick to it. Bare hands melt snow and wet hands will chill quickly by evaporation.

c. Gloves must fit correctly. Gloves must be issued in proper sizes. Tight handwear cuts off the circulation and causes the hands to become cold. To check the size accurately it is helpful to use templates when issuing gloves.

d. At times the fingers become cold and numb, even if all the mitten layers are used properly. A good way to warm them is to swing the arms from the shoulders in a vertical circle, forcing blood back into the finger tips. Another excellent method is to withdraw the arms from the parka sleeves and place the hands under the armpits beneath the parka.

39. MITTENS, SHELL, TRIGGER FINGER. Shell mittens with trigger fingers are designed to be worn over trigger finger insert mittens (see

MITTEN SHELL, TRIGGER FINGER   MITTEN INSERT, TRIGGER FINGER   MITTEN, OVER, WHITE

Figure 13. Handwear.

fig. 13). Two types of both the shell and insert have been issued, some having the trigger finger on the front and others having it on the back of the hand. The mittens fit better when the trigger fingers are on the same side of both inserts and shells; but if they are not, the trigger fingers come nearly to the same position if the inserts are worn on the opposite hand. In windy but not too cold weather, to keep the hands from perspiring, it is often desirable to wear the shells without the woolen inserts. Both mittens are large enough to allow the first finger to be withdrawn from the trigger finger of the mitten and kept with the other fingers for warmth. The thumb, too, may be withdrawn and warmed by the other fingers. When the trigger finger is not in use, it may be turned out and slipped inside the mitten if it is not likely to be needed in a hurry. The gauntlet wrist fits over all the sleeves and should be tightened slightly by means of the strap to insure a good closure. The wriststrap gives adjustability at the wrist.

40. MITTENS, INSERT, TRIGGER FINGER. Woolen insert mittens with trigger fingers are used inside the shell mittens (see fig. 13). They should never be worn alone, because they wear out quickly and catch loose snow. Both pairs of inserts issued to each man should be carried, so that when a pair becomes damp or perspired it may always be replaced by a dry pair. Damp mittens lose their insulating power. Ordinarily inserts should be taken off before the hands get too warm. After a damp pair of mittens is removed, it may be placed inside well-ventilated clothing, or hung securely from the pack to dry. At night, the soldier should take them into his sleeping bag with him to help them dry. Wash insert mittens in lukewarm (never hot) water with soap. Squeeze, but do not rub them, while washing. If these instructions are followed, the inserts will not shrink or lose their shape.

41. GLOVES, SHELL, LEATHER. a. Purpose. Leather glove shells are used to do delicate work with the fingers. Do not use them for heavy labor such as construction or stevedore work. Heavy leather gloves are provided for these purposes.

   b. Fit. This item comes in three sizes: small, medium, and large. The glove shells should be fitted over wool glove inserts so that they avoid sloppiness but do not constrict the hand.

   c. Use. In cold weather, leather glove shells are worn over wool glove inserts. In warmer weather, the shell is used alone to keep the hands from perspiring. In either case, the straps on the back of the wrists should be pulled up to give them a better fit.

   d. Drying and cleaning. Dry these gloves in a moderately warm place away from excessive heat, which will make the leather stiff and reduce their length of life. If the leather does get stiff, work it carefully with the fingers until it is soft again. The gloves may be washed in soap and water, or dry-cleaned.

42. GLOVES, INSERT, WOOL. a. Wool glove inserts are *always* used *inside* leather glove shells. If used alone, they will wear out quickly and, in snowy country, the snow will cling to them and they are likely to get wet. Like the trigger finger insert mitten, they should be removed before the hand gets warm enough to dampen them with perspiration. They are made to fit either hand. They last considerably longer when they are worn on one hand one day, and on the other hand the next day. When only one new glove insert is needed, a single one should be issued as replacement.

   b. For drying, washing, etc., see paragraph 40.

43. MITTENS, OVER, WHITE. White overmittens are used to complete camouflage on snow (see fig. 13). Since their usefulness depends on their whiteness, they must be used only when actually needed for camouflage. When not in use, they should be dried, turned inside out, and stored in a clean place. White overmittens fit over trigger finger shell mittens. There are two types: one with a slit across the whole back of the hand which fits best with the shells that have the trigger finger on the front; and the other with a smaller slit for the trigger finger alone, which may be worn with either type of shell.

Section VI. HEADGEAR

44. GENERAL. a. With proper headgear, it will not be difficult to keep the head and face warm, except in a wind. In very cold weather, the hood of a windproof parka, worn over either a pile parka hood or some warm type of cap, is generally adequate. A face-mask (see fig. 15) is needed by men who must face directly into a cold wind, as in open vehicles and tanks. In extremely low temperatures, especially if the wind is blowing, ears, nose, cheeks and chin are susceptible to frostbite. Shielding the face from the wind with a mitten will help. Making faces will warm the face a little and help to prevent frostbite. Keep a careful watch on the faces of others for the tell-tale sickly white patches indicative of frostbite.

   b. For parka hood, see Parka, Field, Cotton OD (par. 13).

45. CAP, FIELD, PILE, OD. a. Purpose. The pile cap will keep the head warm under extremely cold conditions (see fig. 14). It is particularly useful for vehicle drivers, who need protection against the cold, but must be able to look behind them more easily than is possible in a parka. The cap may be worn under the steel helmet if the headband of the helmet liner is adjusted.

   b. Use. This item has the following adjustable features:

(1) *Visor.* The visor may be pulled down or worn up flat against the cap. It stays in whichever position it is placed.

(2) *Earflaps.* The earflaps may be worn either down around the ears and chin, or tied up on top of the cap. When worn under the chin, the left earflap is slipped through the loop of the right earflap, and tied with a bowknot, overlapping (see fig. 14). When it is not cold enough to wear the earflaps, the left earflap is shoved through the loop of the right earflap and tied on top of the head. If it gets too warm, the flap may be tied up shorter to pull the cap off the ears.

46. CAP, FIELD, COTTON, OD, w/VISOR. The field cap is useful in both hot and cold weather. The visor keeps the sun out of the eyes.

The earflaps are pulled down to protect the ears in cold weather and are turned up inside the cap when the weather is warm. The field cap can usually be worn under the steel helmet without readjusting the helmet-liner headband.

47. CAP, SKI*. Similar to Cap, Field, Cotton, OD (par. 46).

48. TOQUE, WOOL KNIT, M1941 (WITH HOOD CLOTH)*. The wool knit toque is often used under parka hoods in very cold weather. With the cloth hood, it may be worn alone. It covers the ears, chin, cheeks, and forehead, leaving only the mouth, nose and eyes exposed. If worn

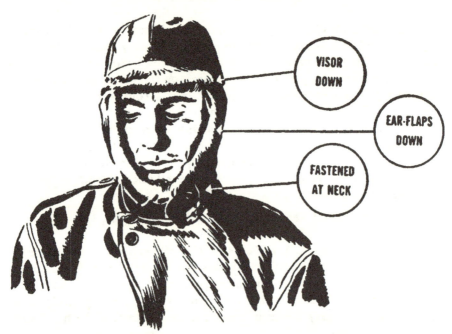

Figure 14. Pile field cap with earflaps and visor down.

across the nose and mouth the moisture in the breath freezes on the toque and may cause frostbite. The toque should be removed if the wearer begins to perspire, for perspiration leads to chilling. The toque may be worn under the steel helmet.

49. CAP, WOOL, KNIT, M1941*. The wool knit cap is used especially under parka hoods, or alone, in windless weather. It has a short visor, and earflaps which may be pulled down when needed. It fits under the steel helmet.

50. HELMET, COMBAT, WINTER. The winter combat helmet is issued to members of the armored force. It is a close-fitting windproof helmet with a woolen lining. It may be washed like any woolen material.

51. HELMET, STEEL, M1 (COMPLETE). Where camouflage on snow is needed, steel helmets (and other equipment) should be covered with special white camouflage paint, which is soluble in gasoline. If the helmet has not been painted, pour water over the outside of it before use, and dip it in the snow. If the snow is partly melted first, it can be put on in lumps which break the outline. Be careful not to touch the helmet with bare hands in cold weather, or skin will stick to it.

52. MASK, FIELD, CHAMOIS. a. Purpose. The face mask (fig. 15) is intended for men who must face winds at cold temperatures. It is particularly useful for vehicle drivers who cannot afford to reduce visibility by wearing the parka hood, or who are facing directly into the wind.

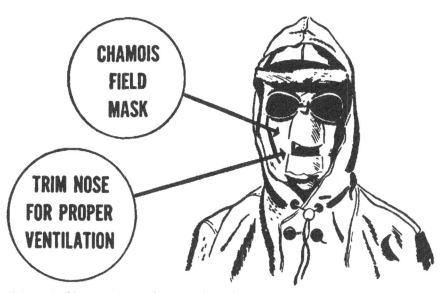

Figure 15. Ski-mountain goggles must always be worn on sunny or bright cloudy days. The chamois field mask is fitted to the individual. Here the nosepiece has been trimmed.

b. Fitting. A new face mask should be carefully fitted to the wearer, in the following manner:

(1) Place the face mask over the face *with the eyeholes still uncut.*

(2) Adjust the strap which runs around the back of the head. It must be tight enough to keep it below the bump at the back of the head.

(3) Adjust the strap which runs over the top of the head, but do not tighten it enough to pull the other strap off the back of the head.

(4) After the straps are perfectly fitted, trace the eyeholes with a pencil. Remove the face mask and cut them out.

(5) Trim the nosepiece if it hangs too low.

c. Use. When the mask is used in extremely cold temperatures, ice will form on it around the nose and mouth. Men who must face a cold wind, as do the drivers of open vehicles, should be provided with several masks, so that they can change at least every hour.

53. GOGGLES, SKI-MOUNTAIN. a. Purpose. Goggles (see fig. 15) are a protection for the eyes against direct and reflected rays of the sun. They must be worn on snow, and occasionally, especially at high altitudes, on light-colored rocks. One hour on bright snow without goggles can cause snow-blindness. Because the eyes are particularly sensitive to diffused light, snow-blindness often occurs on cloudy days when the sun is shining through fog or clouds. Therefore, *goggles must be worn on bright, cloudy days, as well as on sunny days.*

b. Fitting. Goggles can be adjusted to fit the head very comfortably. The head band must not be too tight. If the leather strap on the nose bridge is too long after it has been carefully fitted, the ends may be cut off with enough to spare to keep them from slipping through the buckle.

c. Precautions against fogging. In cold weather, the difference in temperature between the sides of the glass may cause the goggles to fog. This may be remedied by holding them slightly away from the face for a few seconds and letting air get in behind the glass. Paraffin or soap, rubbed on and polished with a soft cloth or an application of gylcerin, helps to keep goggles from fogging in a snowstorm. If the goggles still fog badly, trim the padding around the ventilation holes and punch holes in the fabric. Breathing on goggles in cold weather will cause moisture to freeze on them.

d. Care when not in use. Goggles must be carried in the case when they are not in use; they must not merely be worn on the cap or hung around the neck. Sun reflected from the lenses will give the wearer's presence away to the enemy at a very great distance. Although such reflection cannot be avoided when goggles have to be worn, there is no reason for making oneself needlessly conspicuous when not wearing them. The case will also protect the lenses from getting scratched.

54. HOOD, JACKET, FIELD, M1943. a. General. This detachable hood makes it possible to use the M1943 field jacket at lower temperatures. However, in extremely cold weather, the parka should still be worn. The hood may also be used with the overcoat, field, officer's trench coat type. It may be worn under the steel helmet.

b. Use. (1) Unbutton the shoulder straps, button the hood onto the shoulder strap buttons, rebutton the shoulder straps.

(2) Button the flap of the field jacket under the chin.

(3) Button the right flap onto the top jacket button (the middle of the three exposed buttons). Button the left flap over the right flap

(4) Adjust the drawstring.

CHAPTER 2

EQUIPMENT

Section I. PACKING EQUIPMENT

55. PACK, FIELD. **a.** *Purpose and description.* The field pack is designed to carry the normal load of the infantryman. Other types of specialized packs are issued to each soldier only for certain operations such as those in mountains. The main pouch of the pack is cylindrical in shape, and is covered by a large flap which has a separate, zipper-opening pocket. There are horizontal and vertical straps running around the pack, making it adjustable to the size of the load.

b. *Packing.* (1) It is impossible to give exact packing instructions, for the load differs from time to time and from region to region. Ordinarily, however, heavy objects should be placed near the man's back and hard or sharp objects well towards the center, where they will not chafe the back or rub against the pack. The pack must be kept as compact as possible. Clothing should be folded, or rolled and tied with a thong. When the load is packed it should be well balanced.

(2) *Main pouch.* Objects which are not needed on the march should be packed in the main pouch. If a bag, preferably a waterproof one, is available, it is advisable to place in it everything which will not be needed in combat, so that the whole bag can be removed quickly when the pack is stripped preparatory to going into action. After the pouch is filled, pull the drawstring up tight and tie it with an easily untied knot, taking care to cross-loop the lacing, as shown in figure 16.

c. *Adjusting size of pack.* After the pack is filled, the adjustment straps which run horizontally and vertically around the pack are tightened to make the load as compact as possible (see fig. 17). The two vertical

Figure 16. The drawstring of the field pack is cross looped before being tied.

straps hold down the cover flap and enable the soldier to decrease the length of the pack; the horizontal straps control its diameter. Buckles must always be kept fastened, to keep them from rattling.

(1) *Short pack* (see fig. 17). The field pack is shortened in the following manner:

(*a*) Unhook the bottom ends of the shoulderstraps from the lower set of D-rings.

(*b*) Pull the vertical adjustment straps as far as possible through the slide buckles, which are normally located on the front of the pack about 6 inches above the bottom. Roll the bottom of the pack neatly as the adjustment straps are tightened.

(*c*) Snap the shoulderstraps into the upper set of D-rings.

Figure 17. Pack, Field.

(2) *Combat pack.* When the main pack is empty or nearly empty, the combat pack (see fig. 17) is adjusted in the following manner:

(*a*) Same as for short pack.

(*b*) Same as for short pack.

(*c*) Pull the flap down so that the flap-buckles rest on the upper set of D-rings. Snap the shoulderstraps into both the flap-buckles and the D-rings.

d. Attaching equipment to outside of pack. (1) *Cartridge belt.* The cartridge belt is snapped onto the field pack by means of snaphooks. The snaphooks, which are suspended from the shoulderstraps, fasten to the front of the belt; the hooks which hang on straps attached to the middle of the back of the pack snap into the rear of the belt. In either hot or cold climates, whenever it is necessary to ventilate the body to avoid or evaporate perspiration, the cartridge belt can be left unbuckled and suspended from the shoulders. When no cartridge belt is carried, the snaphook may be snapped into D-rings.

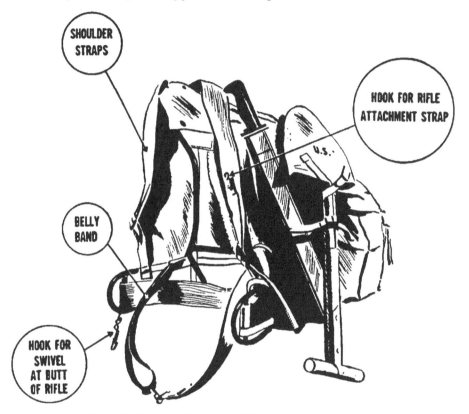

Figure 18. Rucksack from rear. (Note arrangement of straps.)

(2) *Entrenching tool.* The entrenching tool is attached to the center tab on the flap and is carried in the special Carrier, Entrenching, Rucksack. The handle is held under the horizontal adjustment straps.

(3) *Bayonet.* The bayonet is hooked into the tab on the cover flap near the left shoulder, and slipped under the loop halfway down the pack.

e. Camouflage. A tape is sewed in loops to the top of the pack, to hold grass or leaves for camouflage. The horizontal adjustment straps may also be used for this purpose. The vegetation used should be chosen from the commonest kinds in the region. The lighter-colored undersides of the leaves must not be exposed. Short grass and small leaves are usually preferable to large ones since they are less noticeable when the soldier moves.

56. RUCKSACK. a. Nature and purpose. The complete rucksack (see fig. 18) consists of a canvas bag with a map pocket, in the flap which covers the main sack, and three outside pockets; a metal tubing pack frame; a webbing backstrap; a belly strap; and a white camouflage cover. Although the pack board is preferable for heavy loads (over 40 pounds), medium-weight loads (20 to 40 pounds) are carried much more easily in the rucksack than in an ordinary pack. The shoulders take only a part of the weight, for much of it is distributed onto the hips by the frame. The frame also holds the load off the back; allowing air to circulate between the clothes and the load will reduce the amount of perspiration which will collect on the soldier's back.

b. Nature and attachment of parts. (1) *Metal tubing pack frame.* The metal tubing pack frame is easily attached to the canvas bag. The top of the frame is pushed into the leather pocket and held in place by a snap. The bottom of the bag is next strapped tightly to the horns of the pack frame (those parts of the tubing which extend forward around the hips. The strap from the bottom of each side of the pack is run outside and below the metal tubing and back through the D-shaped part of the horn, and then fastened to the buckle located near the strap on the bag. The shoulderstraps are next attached to the frame, behind the perpendicular crossbar of the D. If the pack tends to pull away from the back, these straps may be lashed, with thongs, farther toward the center of the pack along the same piece of tubing.

(2) *Webbing backstrap.* The webbing backstrap runs double between the straight sides of the D of the horns. The flat portion rests on the back, while the buckles are away from it. There are two attachments which hold the backstrap onto a crossbar of the metal pack-frame.

(3) *Belly strap.* The loops for the belly strap are located about three-quarters of the way down the canvas bag in the webbing which runs around the whole pack; at these points, the webbing is not sewed tightly to the canvas. Under rare circumstances, it is more comfortable to run the belly strap doubled between the horns than around the waist.

(4) *White camouflage cover.* The white camouflage cover should be kept scrupulously clean. It is placed over the whole pack, with the drawstrings on both sides. The drawstrings are pulled up and fastened with bowknots. The tie string on the upper side of the cover is passed under the loops connecting the shoulderstraps to the bag, and tied with a bowknot on the leather between the shoulderstraps. The camouflage does not hide the horns or the shoulderstraps. Since the cover would get dirty if always used, it is left behind except when there is snow on the ground.

c. **Precautions in use.** (1) *Load.* Since conditions are different for every operation, the loads carried also differ. Only what is needed is packed. Although it is imperative to figure on most contingencies, the rucksack must not be too heavy. For instance, tent poles and stakes may safely be left behind in wooded terrain, and ski-repair equipment may be eliminated from the packs of units not on skis; but if steep ice slopes are to be climbed, it is foolish not to carry crampons and ice-axes. Obviously, the less that is carried the more mobile a group will be, provided it has everything it needs to be really efficient.

(2) *Manner of carrying.* Unlike the packboard, the rucksack is not designed for carrying the load high. The shoulder straps are left long enough to allow the webbing backstrap to rest comfortably on the back just above the hips. The shoulders get less tired when the arms are being used energetically; for example, when cutting steps on ice, or poling on skis. The webbing backstrap is most comfortable if it is tightened only enough to keep the frame off the back. If the top of the rucksack tends to pull away from the shoulder, the bottom ends of the shoulderstraps may be lashed nearer the center of the frame as described above. The belly strap is fastened in front of the body to keep the rucksack from flopping.

d. **Packing rucksack.** (1) *General procedure.* As a rule, the rucksack is packed with the heaviest objects near the frame. Sharp and hard objects are placed well inside, where they will not rub against the bag. When the load is completely stowed, it must be well balanced. Articles which are needed frequently are carried in the outside pockets or at the top. Maps and other objects which should remain flat are kept in the map pocket. Since everything has to fit into the smallest possible space, clothes are carefully folded or rolled up and tied with thongs, the stove is placed inside the pots of the mountain cook set, and small objects are carried in the canteen cup.

(2) *Suggested arrangement.* Although there are no definite instructions about packing the rucksack, the following suggestions are made:

(a) *In the middle outer pocket:* The 1-quart fuel container. The flap is buckled down.

(b) *In the side outer pockets:* Ski wax, avalanche cord, thongs, sunburn preventive, and other small objects which fit into the canteen cup; extra socks, mittens, insoles, ski climbers, pitons, or any other small items needed on the march.

e. **Fastening loaded pack.** After the rucksack is filled the straps which fasten down the big flap should be pulled up tight and buckled. Because they go around the whole pack, when pulled tight they reduce the size of the pack to a minimum and keep it from flopping. All buckles must be fastened, to keep them from rattling.

f. **Attaching equipment to outside of rucksack.** (1) *General.* Some additional equipment is carried on the outside of the rucksack on the left side. The bayonet is hooked into the flap nearest the left shoulder, while the entrenching tool (in the Carrier, Entrenching, Rucksack), pick, or ax hangs from the middle flap. Both are slipped under the loops halfway down the pack. The first-aid packet is carried on the rearmost flap.

(2) *Rifle.* The rifle is slung from the right side of the rucksack, balancing the above-mentioned objects on the left. It is attached in the following manner:

(*a*) Hold the rifle at the balance point with the left hand, with sling down.

(*b*) Snap the wire hook, which is on the right horn of the pack frame, into the butt swivel on the side toward the body. Hold the piece erect from the swivel.

(*c*) Bring the strap from the right top of the rucksack forward over the right shoulder, passing it outside the piece.

(*d*) Grasp the strap with the left hand and snap it over the head, allowing the rifle to drop between the pack and the right shoulder, with the sling on the body. Place the ring on the strap into the hook on the left shoulderstrap. The ring will slide in and out of the hook most easily if it is turned half over and rolled in or out. The rifle is released by rolling the ring out of the hook, putting the strap over the head, and letting the piece slide down the strap into the hand. The hook is then unsnapped from the butt swivel.

g. **Special uses of rucksack.** (1) *Without frame.* In certain circumstances the rucksack is used without the frame. The frame is frequently discarded for difficult rock climbing, where it would get in the way. The pack is assembled as follows:

(*a*) Unsnap the top of the frame from the leather pocket. Unbuckle the straps from the horns of the frame. Unbuckle the lower ends of the shoulderstraps, and remove them from the frame. The rucksack will then be free from the frame.

(*b*) Rebuckle the straps which were holding the horns of the frame.

(*c*) Slip the end of the shoulderstrap which has the wire loop on it under the horn-strap. Pull the outer end through the loop, thus attaching it to the horn strap.

(*d*) Rebuckle the shoulderstrap and adjust it.

(2) *Prevention of frostbitten toes.* The rucksack may prevent frostbitten toes, if sleeping bags are not available in a bivouac. Remove boots and exchange wet socks for dry ones, then place feet in rucksack, which serves as a windbreak.

(3) *Evacuating casualties.* The rucksack may also be used as an improvised carrying-frame in which to evacuate soldiers with minor injuries, pick-a-back fashion. A strong man can carry a patient in terrain of only moderate difficulty, along a level or downhill, without too much trouble. (For steep climbing, not counting short rises, this method is

much too tiring.) To carry a patient in a rucksack, enough stitches in the seams on both sides of the rucksack should be removed to permit the patient's legs to stick through up to the thigh. He sits in the bag with his legs projecting forward around the hips of the bearer. Clothing is placed in the bottom of the rucksack, so that the patient sits high enough and does not hang down behind. His thighs are held by two or three belts under his knees, strapped together and hung from the bearer's shoulders. He is held to the upper part of the rucksack frame by his own belt, which reinforces the rucksack cord, coming up around his sides. This method can be used for carrying an unconscious patient for he will not be able to tip over backward; but it cannot be used for such serious injuries as broken bones.

(4) For making an emergency sled, see Adaptor, Ski (par. 81).

57. PACKBOARD. a. Purpose. Loads of considerable weight and unusual shape can be carried more easily on a packboard than on any other type of back-packing equipment. Because the weight is distributed over the back and shoulders, 50 pounds can be carried with comparative ease, and loads well over 100 pounds are by no means impossible. The open air space between the body and the load protects the back from hard and irregular objects, and facilitates evaporation of perspiration. This latter function is particularly important in cold weather, as moisture will reduce the insulation of clothing and make it cold.

b. Types. (1) *Yukon type**. A wooden frame with canvas running around the whole board.

(2) *Plywood type.* A bent piece of plywood, with canvas lashed onto the section which rests on the soldier's back (see fig. 19).

c. Preparing packboard for use. The lacings on the canvas of a new packboard are tightened before any load is lashed on. Thereafter the lacings should be tightened whenever they become loose. The lacings of the Yukon type are merely pulled up and tied with a bowknot. The lacings of the plywood type are more complicated. To lace this type, the cord ties into one of the eyelets at the top of the board. Run the cord over the edge, back through the hole of the packboard, down along the packboard, out through the second hole, over the edge of the packboard, down through the second eyelet of the canvas, over the edge of the packboard, back through the hole, etc. After one side is laced and firmly fastened, the other side should be laced as tightly as possible. If the cord is wrapped around a stick each time it comes out from the eyelet in the canvas, it can be pulled much tighter.

d. Packing the load. (1) *General.* Exact instructions for packing a load cannot be given, for the objects carried are always different. It is very important to keep the load compact and *high on the board,* with the heaviest articles on the upper half, and close to the packboard (see fig. 19). Long objects, such as skis, are packed either crosswise or up and down, depending on the terrain. In open country they should be packed horizontally, so they will not swing the packer too much; but in wooded terrain they will usually catch on trees unless carried vertically.

(2) *Packing procedure.* The board is placed flat on the ground, shoulder straps down. When the lashing cord has been undone, a small loop is tied on one end and placed on one of the top hooks. As each article is laid separately on the packboard, the cord is run across and back, as well as up and down in the hooks, and pulled up very tightly. When the whole load is on the board, the cord is fastened with some type of knot which is easy to untie. The loose end is run through

Figure 19. Plywood packboard. The load is kept compact and high on the board.

the uppermost and lowest cross loops, which are then drawn closer together, to tighten the lashing. With a little practice, it becomes fairly easy to pack a neat and secure load which will hold together during an entire day.

e. **Use.** Some loads are so heavy that they cannot be picked up and slung onto the shoulders. With such a load, the packer should set the loaded board upright on the ground, sit down with his back against the canvas, and slip his arms through the shoulder straps. He will then find it easy to roll over onto his hands and knees and get to his feet. Shoulder straps should be kept as short as comfortably possible, so that the load will stay high. If the shoulders get tired, the lower ends of the board behind the hips may be grasped with the hands, which can take the whole weight of the load. When the packer stops to rest, he should try to find a bank or stump on which to rest the load, in order to make it easier to get up again. Regardless of how light the load is, the packer must never let the packboard drop to the ground, or the frame may break.

f. Under sleeping bag. A packboard may be used in place of the insulating sleeping pad under the sleeping bag to protect the soldier's back from snow or cold ground.

g. Care. (1) *General.* A packboard is kept in good repair by mending or replacing a worn canvas before it rips seriously, and by tightening the hooks if they work loose.

(2) *Lashing cord.* When there is no load on the packboard, the lashing cord is wound back and forth around the top of the board (see fig. 19). It is kept on the board and never taken off for any use unless returned immediately.

h. Attachments with quick-release straps. Special attachments and quick-release straps are issued, which greatly increase the efficiency of troops carrying heavy weapons and other awkward loads on the packboard (see fig. 20).

Figure 20. Packboard attachments used with quick-release buckles.

Section II. SLEEPING EQUIPMENT

58. BAG, SLEEPING. a. Use. The sleeping bag is one of the most valuable items of cold-weather equipment. The user must take special care of it, as it is essential to his comfort. The effects of misuse are cumulative, so it is extremely important to learn definite "sleeping technique." An experienced man can sleep with complete comfort in a bag in which an inexperienced person would soon become cold.

(1) *Importance of keeping dry.* A sleeping bag must be kept dry if it is to be warm. This is particularly true in the case of down-filled sleeping bags, for wet down is extremely hard to dry, and lacks the excellent insulating qualities of dry down. Unless proper care is taken in cold weather, a sleeping bag will soon become caked with ice, which makes it frigid. Not only is there danger of the bag's becoming wet from the outside, especially from below, but there is also a greater danger that moisture from the body will condense within.

(2) *Control of perspiration.* Perspiration occurs even in cold weather. It must be held to a minimum to keep the sleeping bag dry. No more clothing should be worn in bed than is needed to keep warm. If the two-layer Arctic bag is issued, the inner case is used alone unless the weather is cold enough for both to be needed. If too much clothing, or too warm a sleeping bag, is used in really cold weather, the sleeper will perspire and find himself inside an inefficient icy casing after a few nights' use. Since some perspiration will always occur, the bag is always opened wide just as soon as the sleeper gets up, and air is pumped in and out to exhaust the moist air and reduce the temperature inside the bag (see fig. 21). Whenever an opportunity arises (during the

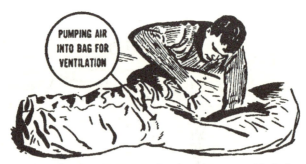

Figure 21. As soon as the sleeper gets up, he should pump cold air into his sleeping bag, and remove the moist, warm air.

noonday halt for instance), the bag should be unrolled and aired. Sun and wind have a drying effect even when the temperature is well below freezing.

(3) *Breathing.* Since a large proportion of the water vapor which escapes continually from the body is exhaled from the mouth, special care is needed to keep the breath from condensing inside the sleeping bag. The head should never be drawn completely into the sleeping bag, away from the face-opening of the hood. In moderately cold weather the nose is always left uncovered in the face opening (see fig. 22). Since the nose is inside this tunnel entrance, it keeps warm enough, but the moist, exhaled air escapes freely and does not condense to any great extent inside the bag. The small frosted area on the hood is easier to dry if the cloth around the edge of the hood is turned in so that the breath strikes on the outside rather than the inside of the hood. In very

cold weather, it is advisable to cover the face with a sweater, muffler, or some such woolen piece of clothing, which will keep the face warm and collect all the moisture, and can be easily dried (see fig. 22).

(4) *Amount of clothing worn.* Experience will help to teach how much clothing to wear in a sleeping bag. It is the tendency of the untrusting beginner to wear too much in a down sleeping bag, and thus to defeat the working principle of the bag. The sleeping bag is intended to be warmed by the person sleeping in it. The warm air in the down provides insulation from the outside cold. Wearing too much

Figure 22. The nose must always be kept in the face opening of the sleeping bag. A wool garment is used to catch moisture from the breath. In cold weather this may be pulled over the face.

clothing, in addition to inducing perspiration, will make the sleeping bag cramped and tight and still further reduce its insulating efficiency.

(5) *Insulation from below.* Added insulation is needed beneath the sleeping bag, especially under the shoulders and hips, because the sleeping bag when it is compressed no longer has the same insulating qualities. If care is not taken in this respect, the soldier's back will get cold, and the heat of his body will melt the snow beneath him. Placing extra clothing underneath the sleeping bag, on top of the insulating pad, is often warmer than wearing it. Packboard, fir boughs, or other materials may be used for added insulation. A down sleeping bag is warmer if fluffed up after being unrolled. This should be a regular practice.

b. Types. There are four types of sleeping bags:

(1) *Bag, Sleeping, Wool.* The wool sleeping bag, which weighs about 3 pounds, has the warmth of about two 4-pound blankets when used without the water-repellent shell, and is even warmer when used with it. It is not intended for really cold weather, but is adequate down to about 40° F. It may be used inside other sleeping bags to give addi-

tional warmth. Like all sleeping bags, it has a mummy shape, a hood, and a quick-release slide fastener.

(2) *Bag, Sleeping, Mountain.* The mountain sleeping bag is a single down and feather filled case, with a hood and a full mummy shape. When properly used, it is adequate for temperatures down to slightly below zero. There are two sizes: regular and large.

(3) *Bag, Sleeping, Arctic.* The Arctic sleeping bag is used in extremely cold weather. It consists of two down and feather filled cases, the inner one being a regular mountain sleeping bag. The outer case has a hood and a full mummy shape, but is closed with snaps. The two cases may be tied together by means of the tapes at the foot of each bag. When the weather is only moderately cold and it is important to cut down on weight, the outer case may be left at the base.

(4) *Bag, Sleeping, Kapok*.* A kapok sleeping bag is issued in some areas.

c. **Case, Water-Repellent.** A water-repellent case is ordinarily used with all sleeping bags, especially in damp or windy climates. In warm weather, when little protection is needed, the case may be used alone. The case fits over the sleeping bag and is laced to the opening. A separate lacing should be used on each side. This lacing should not cross the zipper at any point, and should be knotted at both ends to keep it in place. The snaps are not usually closed, since they merely supplement the zipper, but they provide a good emergency closure. There are slits at the bottom of the case through which the tapes used to tie up the rolled sleeping bag may be slipped. Even with this case, a sleeping bag should be placed, not directly on the snow or wet ground, but on some more waterproof substance, such as a tent floor, a shelter half, or fir boughs.

d. **Liner, Bag, Sleeping.** A washable sleeping bag liner fits all types of sleeping bags. The liner adds somewhat to the warmth of the bag and keeps it a great deal cleaner. It should always be used with the mountain and Arctic sleeping bags, and may be used if desired with the more easily laundered wool sleeping bag. Eyelets at the top of the face opening permit attachment of the liner to all types of bags. The liner is placed inside the sleeping bag, so that the hood fits inside, with the hood flap outside. Laces are then slipped through the eyelets in the liner inside the hood, through the sleeping bag itself, and out through the hood flap of the liner. Tie tapes at the bottom and sides are attached to the corresponding tie tapes in the mountain sleeping bag, to keep it in place. In cold weather, the liner hood flap may be made into the shape of a funnel to catch moisture from the breath, or may be pulled inside the bag to keep drafts from going down past the shoulders.

e. **Quick-release slide fastener.** All newer type sleeping bags are now provided with quick-release slide fasteners, which can be opened in a split second. If closed to about 1 inch from the end of its track, this zipper remains tightly shut; but when pulled to the very end, it opens immediately along its whole length. To close it, run the slider down to the bottom end, thread it into the other side and pull it up again. Be sure that both sides are close together before closing. This feature is ordinarily not used, except in a case of real emergency, for it takes time to rethread the fastener, and it is somewhat difficult to manage with cold fingers.

f. Emergency closure. The sleeping bag has eyelets to which the water-repellent case is usually laced. When the case is on the bag, its snaps provide an emergency closure in case the slide fastener fails. If the bag is being used alone, these eyelets may be used with emergency laces.

g. Use in drying clothing. The sleeping bag is the best place to dry damp clothing at night, when camping outdoors in cold weather. Damp (but not wet) socks, mittens, insoles, etc., are placed in the bag to be dried by the heat of the body. It is wise to take all the socks and insoles which have been worn during the day into the sleeping bag, even if they do not feel damp.

h. Carrying sleeping bags. Sleeping bags may be carried in one of two ways: they may either be stuffed directly into the waterproof sack without being rolled (this helps to keep the down fluffed up), or they may be rolled into a comparatively small bundle and held by the straps provided at the foot for this purpose. They are carried inside their waterproof covers. If the weather is moderately cold, the canteen may be wrapped inside the sleeping bag so that hot liquids will retain their heat and cold ones will be kept from freezing for a few hours.

i. Cleaning. Any type of sleeping bag may be washed. Special instructions have been issued to field laundries on the necessary procedure for washing Arctic and mountain sleeping bags. Wool sleeping bags are washed, like any woolen garment, in lukewarm, never hot, water.

59. PAD, INSULATING, SLEEPING. a. Purpose. The insulating sleeping pad is issued, not to make the sleeping bag softer and more luxurious, but to increase the warmth of the sleeping bag and to keep it dry. The down in a sleeping bag compresses under the body of the sleeper, reducing the amount of entrapped still air and lowering its insulating qualities. As the insulating sleeping pad is not easily compressed, it adds insulation where it is most needed, both to keep the sleeper warm and to prevent the heat of his body from melting the snow beneath. If the cover of the sleeping bag were waterproof, perspiration would condense and wet the bag. The pad therefore performs the second function of keeping moisture from wetting the water-repellent (but never waterproof) bag from below, when camp is pitched on wet ground or snow.

b. Types. (1) *Filled type.* Although the filled type of insulating sleeping pad is not a real air mattress, air must be entrapped among the fibers to give maximum warmth. A small amount of air may therefore be blown into it to keep the fibers from compressing.

(2) *Inflated type.* The inflated type is inflated with the mouth. When there are separate tubes, inflating the outer tubes more than the tubes toward the center will help prevent the sleeper from rolling off the mattress.

c. **Use.** The inflated pad is used under the shoulders and hips. (A rolled-up shirt makes a good pillow. If available, a packboard under the legs and feet is very comfortable; otherwise, clothes or a rucksack complete the insulation beneath the rest of the body.) If either type of pad is inflated too much, it will be very uncomfortable and make the sleeper bounce as he turns over. Before the pad is packed into the rucksack, it should be completely deflated and rolled up to fit into the smallest possible space.

d. **Care.** If the underside of the pad gets wet, it should be aired and dried to prevent its wetting the sleeping bag, and to keep its own weight down.

Section III. COOKING EQUIPMENT AND FOOD BAGS

60. GENERAL. Individual or squad cooking will not be very difficult if well planned in advance. For instance, less time will be lost in the morning if preparations are made the night before. Adequate water may be left in readiness, if it is well insulated from the cold to keep it from freezing. Lunches for the next day may be distributed. Fruit may be stewed for breakfast. The only things then left to be prepared in the morning will be cereal and a beverage.

61. SUPPLY OF WATER. When camping in the snow, water is always a problem. Since it takes about the same amount of heat merely to melt snow as it does to bring cold water to a boil, running water is used, when available, to save fuel. Glacier water is usually clean enough, if the sediment is given a little time to settle. The sun may be used to melt snow. Even when the temperature is low, if snow is sprinkled in a hollow on rocks or on a dark-colored waterproof cloth such as the tent, drops of water will run off and can be caught in a pot. There are many times, however, when it is necessary to melt snow or ice to obtain water. Since ice and wet or frozen snow have a much higher water content than powder snow, they are more satisfactory. It is therefore often worthwhile to scrape off surface powder snow to get to more solid snow for melting. It is easy to burn the pot when powder snow is used, for the snow near the bottom melts first and is drawn up into the drier snow above. Snow must therefore be stirred constantly and shoved to the bottom of the pot until the whole is slushy (see fig. 23). Take great care to use only uncontaminated snow.

62. STOVE, GASOLINE, ONE-BURNER, M1941 AND M1942. One-burner gasoline cooking stoves are issued for the use of mobile troops who prepare and cook their own food. These stoves are easy to manage, burn little fuel, and are light in weight.

a. General precautions in use. (1) *Ventilation.* The combustion of the fuel in these small gasoline stoves is usually almost perfect, and there is little danger from fumes. When the weather is cold, however, or when a pot filled with snow is placed on the stove, the combustion will not be complete, as the gases cool rapidly. As a result, poisonous but odorless

Figure 23. Melting snow in mountain cook set, with M1942 one-burner gasoline stove. The snow must be stirred and shoved to the bottom, to keep the pot from burning.

carbon monoxide gas escapes into the atmosphere. Therefore, the colder the weather, the greater the need for ventilating a tent when the stove is burning. As the mountain tent is absolutely airtight, extreme care must be taken to keep the tent ventilated when a stove is lighted, particularly when cooking. Ventilators must never be tied shut, regardless of the temperature. In addition the door should be left open, if possible.

(2) *Steam.* When liquid is boiling on the stove, the pot should be covered to keep steam from rising, forming as frost on the tent roof, and wetting everything in the tent. Use as low a flame as practicable, to minimize moisture and save fuel.

(3) *Wind.* When the stove is used outdoors, it must be shielded from the wind.

(4) *Fuel.* If small gasoline stoves are filled more than three-quarters full, they flare up and, although they will not explode, the flame may

burn the tent or other equipment. Nonleaded "white" gas should be used when available. Kerosene is quite satisfactory.

(5) *Lubrication.* Pump washers should be oiled regularly.

b. Types. Of the two types of one-burner gasoline cooking stoves, the M1942 is decidedly the lighter and more compact.

(1) *Stove, Gasoline, One-Burner, M1941* (see fig. 24). Before this stove is lighted, the covering cups are removed and the legs and pot arms snapped out. The stove is lighted and cared for in the following manner:

(*a*) Unscrew the pump handle by turning it two turns to the left. Holding the thumb over the venthole in the cup of the pump handle, pump about 30 times. Push the pump handle in and screw it to the right until tight.

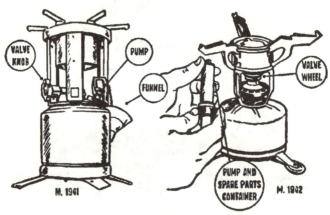

Figure 24. The M1941 and M1942 one-burner gasoline stoves.

(*b*) Revolve the wire lever near the valve knob several times, to clean the tip. Leave the lever in a down position.

(*c*) Light a match.

(*d*) Turn the valve knob a quarter turn to the left (counterclockwise), and light the stove by applying the match to the top of the burner. Do not open the valve farther at this time.

(*e*) After 3 or 4 minutes, when the yellow flame subsides and the flame burns completely blue, turn the valve knob as far to the left as possible. Pump more air into the stove until the flame burns vigorously.

(*f*) If the flame is too hot, turn up the wire-cleaning lever to adjust the heat, and throttle the flame down. If the flame starts to flicker, clean the tip by twisting this lever around several times. Since this will probably put out the flame, keep a lighted match in readiness.

(*g*) When the stove is no longer needed, extinguish the flame by turning the valve knob to the right to close the valve.

(*h*) If the stove is not functioning properly, use the extra generator assembly attached to the inside of one of the uprights. To install the new generator assembly, the crown of the burner is unscrewed counterclockwise. The nut which holds the old generator in place is then unscrewed, and the old generator removed. If the needle which remains exposed is damaged, it is also unscrewed. The nut on the old needle is removed and placed on the new needle, which is then screwed firmly into place. When the cleaning wire lever is up and the new generator is in place, the needle should be exposed about $\frac{1}{16}$ inch through the burner orifice. Then the new generator is bolted down and the rest of the burner screwed back on. If the cleaner lever does not move freely up and down, the cleaner needle has not been screwed down far enough. The needle must be handled carefully so as not to damage it when making repairs. When the stove is packed to be carried with gasoline in it, loosen the screw caps on the gasoline tank to release the pressure, and then retighten. Check the valve to see that it is shut. Place the small utility cups down over the burner carefully, so that they will not jam on the chain.

(2) *Stove, Gasoline, One-Burner, M1942.* This stove is as easy to use as the M1941. Before it is lighted, its legs are snapped out. The pot arms are then raised allowing the ring on which they rest to slide down into position. The stove is lighted and controlled in the following manner:

(a) Before pumping air pressure into the stove, to clean the tip, turn the wheel under the burner to the right until it stops.

(b) Turn the wheel back to the left until it stops, closing the valve.

(c) Pump about 35 strokes.

(d) Light a match.

(e) Open the valve slightly by turning the wheel to the right, and light the stove by applying the match to the top of the burner. Do not open the valve farther at this time.

(f) When the flame burns blue, turn the wheel to the right a full turn.

(g) Pump until the flame burns vigorously. After the stove is burning well, the heat of the flame may be reduced by turning the wheel to the right until the cleaner needle enters the orifice.

(h) If the flame flickers, clean the tip by turning the wheel to the right until it stops. Turn wheel back to the left and relight. Extinguish the flame by turning the wheel to the left until it shuts the valve.

(i) If the cleaning needle does not work properly, or if the flame seems faulty, inspect the generator and needle, and if necessary, replace. This is done in the following manner:

    1. Unscrew the burner head counterclockwise. Shut the valve by turning it to the left, and, holding the valve wheel stationary, loosen the large nut below the priming cap.

    2. Lift the generator out through the top opening.

    3. If the needle requires changing, open the valve by turning it to the right until it stops.

4. Unscrew the needle and remove the small nut. Screw the nut onto the new needle and put it back into position. When the generator is properly seated, the needle should show about $\frac{1}{16}$ inch above the opening.

5. Reassemble.

(j) If the stove leaks below the valve wheel and above the tank connection, tighten the nut below the valve wheel to eliminate the leak. If this is not effective, replace the packing, in the following manner:

1. Loosen the nut below the valve wheel until it drops down. Raise the wheel directly upward and remove it carefully.
2. Unsnap the small wire ring around the exposed portion of the valve, and remove it.
3. Lift the ring off the whole assembly.
4. Remove the packing and the packing rings, and insert the new packing.
5. Reassemble carefully.

(k) If air leaks out of the pump, check to see if it needs a new pump cap gasket. Remove the whole pump from the tank and unscrew the knurled section from the pump body, exposing the gasket. Then remove the old gasket, if necessary, and replace with the new one, which must be packed completely into the circular recess.

(l) If gasoline leaks out of the pump, check to see if it needs a new pump check. If it does, repair as follows:

1. Remove the pump from the tank.
2. Unscrew the nut at the bottom end of the pump. (Be careful not to lose the small spring.)
3. Remove the old check.
4. Insert the spring into the new check.
5. Insert the spring and check into the nut, removing the spring first.
6. Reassemble.

(m) Before it is packed for carrying (in the mountain cook set), cool the whole stove, and check the wheel to see that the valve is closed. Loosen pump cap and tighten it, to let out gasoline vapor, which might otherwise build up pressure tending to cause leakage of gasoline into the cook set. If gasoline should leak into the cook set, heat the empty pots carefully before cooking, to drive off the gasoline.

(3) *Stove, Cooking, Gasoline, M1942 Modified.* The modified M1942 is similar to the older model but has a different type of burner which is somewhat similar in appearance to that of the M1941 model. The stove is lighted and cared for in the following manner:

(a) Clean the tip by moving the cleaner lever down and back up.

(b) Pump 20 to 30 strokes.

(c) Open the valve slightly by turning the valve knob counterclockwise and allow the priming cup to fill three-quarters full. Close the valve.

(d) Light the gasoline in the priming cup.

(e) When the flame has almost died down, open the valve gradually until the maximum blue flame appears.

(f) The flame may be reduced somewhat by lowering the cleaning needle a little.

(g) To extinguish, turn the valve clockwise until the flame is extinguished.

(h) A maintenance wrench is attached to the side of the stove. This may be removed by pulling out the spring clip slightly and by lifting the wrench directly upward.

(i) If the stove is not functioning properly, use the extra generator assembly attached to the inside of one of the uprights. To install the new generator assembly, the crown of the burner is unscrewed counterclockwise. The nut which holds the old generator in place is then unscrewed, and the old generator removed. If the needle, which remains exposed, is damaged, it is also unscrewed. The new needle is then screwed firmly into place. Then the new generator is bolted down and the rest of the burner screwed back on. If the cleaner lever does not move freely up and down, the cleaning needle has not been screwed down far enough. The needle must be handled carefully so as not to damage it when making repairs.

(j) If gasoline leaks from around the cleaning lever, the packing nut should be tightened slightly until the leak stops. If this is not adequate a spare packing is found in the spare parts container attached to one of the uprights. Unscrew the plastic knob on the lever, remove the packing nut, pull out the lever body including the lever packing gland and the old packing. Separate the lever packing gland and the old packing from the lever body. Reinsert the lever body into the stove and fit the packing very carefully into position. (This packing crumbles easily and must be handled with care. If it breaks, it may, in an emergency, be replaced with graphited asbestos cord.) Insert lever packing gland, being certain it is properly seated. Attach the lever packing nut and tighten moderately. Reattach the lever knob.

(k) If gasoline leaks from the valve stem tighten the valve packing nut moderately until the leak stops. If this is not adequate, a spare packing is found in the pump barrel. This should be inserted to replace the old packing.

(l) If the pump leaks, repair as in (2) (k) and (l) above.

c. **Replacement parts.** These parts are carried in the pump plunger. When the pump plunger is pulled out, the pump handle may be unscrewed. In the plunger are the following items: generator assembly (including a cleaning needle, an orifice, screens, and a generator), extra screen, valve packing, pump check and pump cap gasket.

63. COOK SET, MOUNTAIN. a. Nature. The mountain cook set is composed of two pots which nest into each other and a cover for the larger pot (see fig. 23). The cover is also used as a frying pan.

b. Care. The pots should be kept scrupulously clean, so that the user's health will not suffer. Sand or granular snow is useful for scrubbing the pots, which should also be washed in extremely hot water to remove grease. If used over open wood fires, the outsides of the pots become blackened. They must be cleaned before being put into the rucksack to keep clothes from getting dirty.

c. Carrying M1942 gasoline stove. (1) The M1942 gasoline stove is carried in the cook set. The pots are nested together, the stove placed inside, and the cover placed on the top.

(2) When the tactical situation requires silent movement, the stove may be wrapped in a sock to keep it from rattling around.

(3) The air valve must be opened before packing, to prevent gasoline from leaking into the cook set.

64. CONTAINER, FUEL, 1-QUART. a. Purpose. This container is issued for carrying fuel and for filling small gasoline stoves. It fits into the middle outside pocket of the rucksack, where it is carried well away from any rations.

b. Use. (1) Slight seepage of gas may take place during carrying. After unit is unpacked and before using, this gas may be driven off by heating *carefully* over a low flame while the pots are still empty.

(2) There are two types of caps. In one type, a piece of tubing projects into the can from the screw cap, and is screwed firmly into place when the container is being used only to carry fuel. When it is needed to fill a stove, the plug in the screw cap is unscrewed, releasing the tube, which is pulled up until the lower plug stops it. This lower plug is then threaded into the socket. When the tube has been inserted into the fuel opening of the stove, the container is tipped to allow the fuel to enter the tube and run into the stove through the hole near its upper end. Squeezing the sides of the can speeds the flow of gasoline.

(3) With the other type, there is a tube fixed horizontally to the cap. When the washer inside one cap is removed, gasoline can flow freely out of either end.

65. BAG, FOOD, WATERPROOF. a. Purpose and nature. Waterproof food bags are issued principally for carrying food. A complete set consists of four bags, each a different size.

b. Use. As the use for each bag is not definitely specified, the mountain trooper may use his own judgment about what to put in each. For example, a bag may be used to hold half a can of powdered milk, or the remainder of some cereal. It is suggested that one bag be reserved for the food which will be used on the march during the course of the day. The bags may be used for objects other than food which must also be kept dry. The tie string at the top of the bag permits an almost perfect watertight closure. This string may be simply looped twice around the mouth of the bag, pulled up very tight, and tied with a bowknot. However, closure is better if the mouth of the bag is folded over and held firmly in place by the tie string.

c. Care. After use, the bag should be turned inside out, brushed, and aired. Occasionally it should be washed in fairly hot (but never boiling) water.

Section IV. MAKING CAMP

66. TENT, MOUNTAIN, TWO-MAN, COMPLETE. a. (1) The mountain tent (see fig. 25) is a lightweight, waterproof, two-man tent with a floor. It is reversible and may be pitched with either the olive-drab or white side out, depending on which will provide the best camouflage. A complete tent unit consists of the following items:

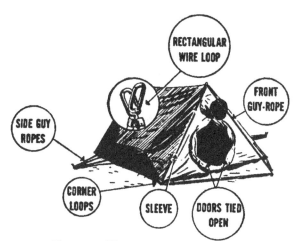

Figure 25. The mountain tent, pitched.

(*a*) One tent.
(*b*) Twelve tent pole sections (four complete tent poles).
(*c*) Six tent pegs.

(2) There are two types of tent poles, each with a different kind of top section. One type has a rectangular wire loop at the top of the pole (see fig. 25). No toggle pin is needed with this type of pole. The other has a circular disk with the eyelet on the end of the top section. Two toggle pins are necessary with this type. They should be attached to the webbing at the top of the ventilators.

b. The complete unit need not always be carried. Snow and bough shelters may be used in place of the mountain tent, either intentionally to save weight, or in an emergency. These are fully described in FM 31-15. Occasionally, in wooded terrain, the tent alone is used in order to save weight and achieve maximum mobility. In such cases the corners are staked down with any sticks or rocks that are available. The front and rear guy ropes are staked out onto sticks or rocks. If the ridge of the tent sags, it may be supported by the loop located at the center of the ridge. Skis and ski-poles may also be used, in place of tent poles and pegs.

67. PITCHING THE TENT. a. Wind consideration. The prevailing wind should be taken into consideration. If possible, the tent should be pitched with one of the back corners into the wind, rather than a side or the front. If the tent is pitched on snow with the entrance directly down wind, the entrance is likely to become blocked, as drifting snow tends to pile up in the lee of any object.

b. General procedure. With a little practice, the tent can be pitched quickly. The normal procedure is as follows: (1) The tent is reversible, and may be turned inside out if necessary. Start with the proper color on the outside.

(2) Divide the tent pole sections into four groups of three so that each has a bottom section with a spiked end, a middle section, and a top section with an eye and disk, or a rectangular wire loop.

(3) Assemble two of the poles in the outside sleeves in the front. The sleeves are found inside and outside the tent, along the seams which join the sides to the front and to the back. Slip the spikes on the bottom of the poles through the loops in the corners.

(a) If the poles have rectangular wire loops at the top, slip one loop through the other. Twist the pole a half-turn, to lock the loops together.

(b) If the poles have eyes and disks at the top, place the eyes over each other. Slip the toggle through both eyes and through the grommet at the peak of the tent. Lock the toggle.

(4) Assemble the rear poles similarly.

(5) Attach the guy ropes to the webbing loops on the front and rear peaks of the tent.

(6) Stake out the front and rear guy ropes on tent pegs.

(7) Attach side guy to loops on the sides of the tent.

(8) Stake out the side guy ropes on tent pegs. Both ropes may be attached to the same peg.

(9) To anchor the corners, tie ropes to the loops at the corners and stake them to the front or back peg.

c. Special procedure. (1) In rocky terrain, it may be impossible to drive the tent pegs into the ground. In that case, attach the guy ropes to rocks.

(2) If the snow on which the tent is pitched is loose and powdery, the guy ropes may be attached to skis, ski-poles or ice axes driven into the snow, after it has been packed.

(3) Ropes from the loops at the corners to the front or back peg will anchor the corners.

68. PERSONAL PRECAUTIONS. a. Care of clothes and boots. Before entering the tent, each man must be very careful to brush all the snow off his clothes and boots. Snow brought into the tent will melt and wet sleeping bags and clothes. If two men enter a tent, one man should come in first and take the bags, packs, etc., from the other, after the latter has brushed them off carefully. No small objects should be left loose outside the tent, for they will sink into the snow and get buried and lost. Care should be used, when wearing nailed boots in a tent, not to rip the floor.

b. Closing the entrance. The tunnel entrance of the tent may be tied up tight by means of the tie string from either the inside or the outside. The tie string is merely wound around the tunnel entrance as if it were the mouth of a bag, and fastened with a half-hitch. The entrance may be kept wide open by tying together the tapes located around the edge of the tunnel (see fig. 25).

c. Ventilation. Ventilation is of the greatest importance in the mountain tent, for the cloth is absolutely impermeable. It is dangerous to close up the tent completely, especially if a stove is burning. To insure a good supply of fresh air at all times and in all kinds of weather, ventilators are provided at each end of the tent. In good weather, they are usually kept wide open by hooking them up to the guy ropes. In most storms, they will provide adequate protection as well as ventilation if left hanging loosely. Only under exceptional circumstances are the ventilators closed with the tapes, and never when a stove is lighted, because of the danger of carbon monoxide. In cold weather, there is an additional reason for leaving them open. Unless the moisture caused by breathing and cooking can pass off into the outside air, it forms as frost inside the roof of the tent. In a wind, this shakes off and wets the clothes and sleeping bags. To keep this from happening, brush frost from the roof, walls, and floor and throw it out.

69. BREAKING CAMP. When breaking camp, if the mountain tent has frozen in as the result of successive thawings and freezings, it must be handled very gently. Shovels must be used with the greatest care, since it is very easy to rip the tent while digging it out. Ice remaining on the tent should be carefully removed before the tent is packed up.

Section V. MOUNTAIN CLIMBING EQUIPMENT

70. AX, ICE, MOUNTAIN. a. Purpose. The ice ax is useful, not only on ice and snow, but in most types of montainous terrain. As a walking stick, it aids balance and helps reduce the weight on the legs, particularly when heavy loads are carried. It is also used for cutting steps on grass and gravel slopes. When the pick end of the head is firmly shoved into turf, gravel, or hard snow, the blade end makes an excellent handhold. In the first World War, the ice ax served at times as a very effective weapon in close-in fighting on the Alpine front. The most important use of the ice ax, however, is on ice and snow. The pick end of the head is essential for chopping steps in ice, while the blade is used to cut steps in hard snow and to clean them of ice chips. The ice ax is needed for belaying the rope on snow and ice to safeguard the climbers. Suspected snow should be probed with the ice ax to see whether there are hidden crevasses under it.

b. Use. (1) *On easy terrain.* In easy terrain when the ice ax is used as a walking stick, the hand grasps the head with the pick forward. The palm rests on the blade of the head, while the sling hangs around the wrist. Sometimes the ice ax is not needed, and sometimes the noise of the point striking rocky ground may make its use inadvisable. In such cases, there are two easy methods of carrying it. The shaft may be held in the hand with the point forward, where it is out of the way of anyone following, or the ax may be held under the arm with its head resting against the back and its shaft held up by the forearm.

(2) *On slopes.* In traversing steep grass, rock, snow, or ice slopes, the ice ax is an indispensable aid to balance. The head of the ax is held in the hand away from the slope, while the shaft, which runs diagonally across in front of the body, is grasped by the hand toward the slope. If the climber slips, he can throw his weight onto the shaft and save himself from falling, since the point will catch on the slope. It is used similarly for glissading or sliding down snow or scree slopes. For climbing up snow it is usually driven in vertically.

(3) *On rock.* On rock, the ice ax is more of a problem. If not needed, it may hang from the wrist on the sling. For short pitches, it is sometimes tucked into the belt or rope, but care is needed to keep it from falling out. It may be shoved between the wearer's back and the rucksack, with its point down and its head on the top of the pack. For long rock climbs, however, it is usually packed in the rucksack. The head rests on the bottom of the bag and the point sticks straight up out of the pack.

(4) *On snow and ice.* (a) *Cutting steps.*
    1. Step cutting needs a great deal of practice. With both hands on the shaft, one through the wrist sling, the climber strikes two or three blows with the pick across the ice in a straight

line perpendicular to the slope. He then strikes at a slight downward angle into the ice above this line, finishing the step. If the ice is brittle, he must be careful not to strike too hard and chip the whole surface away. On hard snow the blade end of the head can be used to scrape out a step in a single blow.
2. Ice steps need be only big enough so that the foot can rest safely on them. Their surface should slope in slightly. Since it is easiest to chop steps diagonally going up hill, a big ice slope will be ascended in zigzags, with larger steps at the turns. While chopping, the climber stands with his outside foot in the lower step, and may balance himself with his uphill knee which rests against the slope. Ordinarily, two steps are cut from one position.
3. On very steep slopes, handholds are sometimes chopped.
4. Cutting down hill is tiring and difficult, since the climber must lean forward down the slope.
5. The use of crampons eliminates the need for step-cutting on moderate slopes, and makes climbing ice steps on steep slopes much safer.

(b) *Belaying.* The ice ax is used for belaying the rope on snow and ice to safeguard the climbers. Two methods are described below:
1. A shaft belay is used, unless the snow or ice is so hard that most of the shaft of the ice ax cannot be driven in. It is well to steady the ax with the hand and knee. The rope is looped around the shaft near the surface of the snow, and both strands are held together in the hand. The rope is taken in or payed out to keep it taut as the climber moves.
2. When the ice is too solid for the shaft belay, the pick belay is used. The pick is driven into the ice as far as possible. While one hand or knee keeps it solidly in the ice, the other pays the rope over the head of the ax.

c. **Care.** (1) *General.* Since the life of a mountain trooper often depends on his ice ax, he must take good care of the ax. The shaft is treated with linseed oil two or three times a year. If it gets rough, it should be sandpapered smooth to save the hands. A weakened shaft must be replaced at the first opportunity. Whenever the point on the bottom of the ice ax gets dull, it should be filed sharp again or it will slip. A bit of oil on the metal parts of the ice ax when it is not in use will keep it from rusting.

(2) *During storms.* Mountain troopers must often operate in storms. Lightning is especially dangerous in the mountains. Since ice axes, like all metal equipment, attract lightning, they should be laid aside while the climbers wait at some distance from the ridge top.

71. CRAMPONS, MOUNTAIN. a. Purpose and nature. The soles of unnailed boots are dangerously slippery on hard ice even where a mountain slope is only moderately steep, whether or not steps are chopped in the ice. The nails or cleated soles of ski-mountain boots help hold on such a slope, and may be used in ice steps; but they, too, have their limitations on hard ice and steep slopes. Therefore crampons (see fig. 26) are often used, to give the feet a safer grip on icy slopes. Crampons save time and energy as well, for they often eliminate step cutting. Chopping steps is time consuming and tiring. Where a group without crampons must work its way laboriously and slowly up hill, another group with

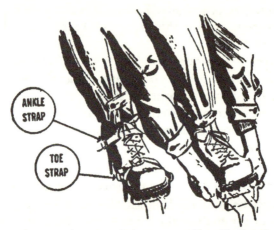

Figure 26. Crampons being put on. (Note bindings.)

crampons can move rapidly and with much less effort. Even with crampons, the ice may sometimes be so hard and the slope so steep that steps have to be cut; but much smaller steps are sufficient. Crampons make it possible to climb almost noiselessly, eliminating the sound of the ice ax striking the ice and the clatter of ice chips. Crampons are also very useful on extremely steep grass slopes.

b. Fitting. Crampons are provided in two sizes because they must fit well on the soles of the boots. Since the length is not adjustable, the proper size should be chosen when the crampons are issued. The front points should be under the very toe of the boot, when the foot rests in position on the crampon. The width is easily adjusted to fit the sole exactly, by spreading or narrowing the uprights of the crampons with a hammer, preferably in a vise. The fit is correct when all the points are under the edge of the boot, and when the upright binding pieces fit snugly around the sole.

c. Attaching. To attach the crampon to the boot, it is placed on the snow with the points down, with the metal band connecting the two rear rings toward the heel, and the buckles outside. The foot is placed on the crampon and pushed back snugly against the heel band, which is raised to grasp the boot firmly. The rear straps are buckled over the ankle and the front straps fastened across the foot (see fig. 26). These straps must be buckled so tightly that they will not allow the crampon any play on the foot, but not so tightly that they will cut off the circulation and cause frostbite. If a binding breaks when in use, a long thong or strap may be substituted. This is threaded from the rear outside ring, to the middle inside ring, through the front outside ring, through the front inside ring, through the middle outside ring, through the rear inside ring, and then tied across the ankle.

d. Use. The use of crampons requires a certain amount of practice. When wearing crampons, the foot should not be edged into the slope. On the steep slope, the ankle must be bent considerably to allow the foot to strike flat on the slope with all points biting into the ice. The climber should swing his foot forward without bringing it close to the other leg. Trousers snagged on crampons points have caused serious falls. If there is loose snow on the ice, it may ball up in the crampon and keep the points from gripping on the hard surface below. When this happens, the foot should be raised and the edge of the boot struck with the shaft of the ice ax. Where there are short stretches of rock, it is usually not worthwhile to remove the crampons. Care is needed, however, since they do not grip well on rock.

e. Care. Crampons should be kept free from rust. If they are not in frequent use, a little oil now and then will keep the metal in good condition. When the points become dull, from use on rocks, for instance, they should be filed sharp again. It is advisable to inspect crampons occasionally for splits or cracks in the bends of the metal. If a point does break off on a steep slope, a bad fall is almost sure to result. The binding will last much longer if dubbin is applied occasionally. The amount of dubbin used should be limited, however, otherwise the straps will get stiff in cold weather. A weak binding must be replaced at the first opportunity.

f. Placing in rucksack. Crampons are rather a nuisance in the rucksack, since clothing, tents, and sleeping bags are easily snagged on the points. They are held by the binding straps with their points together and placed under the large flap on the top of the rucksack. They should not be hung on the outside or they might flop and be noisy.

72. ROPES, CLIMBING. *a. Purpose.* Climbing rope is used to safeguard the lives of mountain troopers when climbing, and to help them carry out many of their duties. Although its principal use is in safeguarding climbers on difficult mountain slopes, it also serves for lashing loads down, for hauling up or lowering loads, for making stirrups with which to climb out of crevasses or overhangs, for improvising rope ladders, etc. The larger $7/16$ inch rope is always used for roping men together and may also be used for the other purposes mentioned above. The

¼ inch rope is intended only for the auxiliary uses, and to serve as a fixed rope.

b. **Treatment of new rope.** Because a new rope kinks easily, it is worthwhile working a rope over carefully when new. It is not considered advisable to wet and stretch a new rope, as this destroys essential resilience. The middle, and perhaps also the quarters, of a rope should be clearly marked with some simple identification, such as colored thread woven through the strands.

c. **Care in use.** (1) Ropes fray when dragged under tension over sharp rock edges. If a slip occurs when the rope is running across such an edge, it may break at that point. Therefore if it is impossible to avoid sharp edges, hammer them off or pad them with clothing or some other material. At the end of a day of rock climbing, the rope should be examined carefully along its whole length to see if it has become snagged or cut. Nylon rope frays slightly on the surface without causing any real loss of strength. Be careful to see whether the rope is frayed only on the surface or is really snagged. Loose ends must be rewhipped immediately. The ends of a nylon rope may be fused in a hot flame in addition to whipping.

(2) The rope should be kept taut between men so it will not get wet from dragging on snow. Wet rope should be dried completely, at the first opportunity, by uncoiling it and hanging it in big, loose loops, in a cool, dry place where air is circulating.

d. **Precautions when rope is not in use.** The easiest way to prevent the rope from tangling when not in use is to keep it well coiled. The right size coil will result if the climber sits down and coils the rope over his knee and foot, after first securing one end with a square knot. He coils it until about 8 feet are left. After fastening these with a half-hitch, he winds the remaining end tightly around and around the whole coil until it is used up. The very end is secured by shoving it into the coil or by tying it. Climbing rope loses much of its strength when stored and not used. It deteriorates least when kept in a cool, dry place. The rope should be kept out of strong sunlight when not in use.

73. PITONS, MOUNTAIN; SNAPLINKS, AND HAMMER PITON. a. Purpose. Pitons are driven in with a piton hammer and a snaplink is always snapped into the ring of the piton to hold the rope (see fig. 27). Pitons are driven into cracks in rocks and into ice as safeguards to difficult climbing where there is no projection of rock; pitons and rope slings may hold a doubled rope for roping down. They may also be made into improvised pulleys for hauling heavy loads up precipitous slopes.

b. **Types.** (1) *Pitons, Rock.* Pitons are driven in with the ring on the lower side. Some cracks run horizontally across a cliff, while others run vertically up and down. To keep the ring on the lower side, there are therefore both *horizontal and vertical* pitons (types I and II).

Sometimes such pitons cannot be driven in with rings in position to hold the snaplink. In such a case, a *wafer rock piton* (type III), with a movable ring, is preferable. In wide cracks or in comparatively soft rock, *angle rock pitons* (type V) are used. Since it is obvious that pitons will have more of a tendency to pull out of vertical cracks than horizontal ones, greater care must be taken to see that vertical pitons are firmly driven into the rocks. Pitons driven in downward may hold well, but if they are driven in upward, they are rarely much of a safeguard. A secure piton will sing on a characteristic high note when it

Figure 27. A piton being driven in with a piton hammer. The rope runs through the snaplink which hangs from a second piton (lower right). The climber is thus safely belayed.

is driven well home. Each piton should be tested before relying upon it. When the stance of the climber is rather difficult, avoid loss of pitons by snapping them into a snaplink hanging on a thong from the neck, while driving them in. Pitons are pulled out by the last climber whenever possible, and used again. Old pitons found in the rocks must always be carefully tested; usually they should be removed and redriven.

(2) *Pitons, Ice.* There is only one type of ice piton (type IV) issued to mountain troops; it is a tubular type with a movable ring. Although most of the displaced ice goes up the tube when it is driven in, there is always danger that the force of the blows will cause brittle ice to flake off. Before pitons are driven, rotten surface ice should be chipped

off until sound ice is reached. Ice pitons are usually driven in slightly downward from a line perpendicular to the slope until all but about 1 inch of the tube is in the ice. Fixed ropes cannot be left attached to ice pitons for more than a short time, for the piton is likely to melt out. When the temperature is near freezing or higher, the tension on the piton may lower the melting point of the ice so that the piton becomes insecure. When the sun is strong, ice pitons should be placed in the shade, if practicable. Like rock pitons, ice pitons should, whenever possible, be removed and used again.

c. **Snaplinks.** Snaplinks are always used with pitons, since otherwise it would be necessary to unrope in order to thread the rope through the piton. After the piton is driven in, the snaplink is snapped into the ring of the piton and then around the rope (see fig. 27). Care is needed to keep the movable lever away from the rock, so that the snaplink will not open from pressure against the rock. When climbing with pitons, the rope leader carries several snaplinks on his belt or on the noose of rope around his waist, where he can get at them easily. If they are on the noose, he must be careful when unroping to keep them from falling and getting lost. When the last man on the rope reaches the snaplink, he unsnaps it from the piton, but leaves it on the rope until he is in a position to snap it onto his belt or to return it to the rope leader. Snaplinks are never left behind in pitons.

d. **Hammer, Piton, Mountain.** (1) *Nature and use.* The head of the piton hammer (see fig. 27) has a flat striking surface on one end, and a pick on the other. The former is used principally for driving and removing pitons. It is also useful for smoothing off sharp edges and jagged rocks, which otherwise would fray the rope. The pick is used for chopping corners and insecure rocks out of cracks. It may also be used for scraping ice glaze off rocks. On ice it makes a poor substitute for an ice ax. The head of the hammer must never be used as a lever.

(2) *Manner of carrying.* The piton hammer may be hung from the wrist on a leather thong at the end of the short hickory handle. The ends of the thong are fastened by means of two slits in the leather. The slit on one end is opened and the other end slipped through it. The slit on the second is then opened, and the middle of the first strap is pulled through to make an adjustable loop for the wrist. The thong is adjusted to the wrist. Some climbers prefer the longer emergency thong which can be worn around the neck leaving the hands free.

Section VI. SKIING EQUIPMENT

74. GENERAL. For operations in deep snow, skis or snowshoes are needed. Skis are much faster, especially down hill, and less tiring to soldiers skilled in their use. However, for pulling sleds, carrying very heavy loads, or operating in thick woods, snowshoes are preferable.

A soldier on skis needs a considerable amount of equipment aside from the skis themselves. Unless a flexible ski binding is available, he must wear ski-mountain boots, well fitted to the ski bindings. He always carries a pair of ski-poles. Ski wax is essential for sliding down hill and for climbing up hill, as well as to keep the skis in good condition. Mohair ski climbers save a great deal of energy on long climbs. When skis are damaged in the field, emergency ski tips, contraction bands, and a ski repair kit are needed to make temporary repairs. A soldier on skis can improvise a sled from his equipment, if he carries ski adaptors.

75. SKIS. a. Nature and fitting. Both laminated and solid flat-top hickory skis are provided in four lengths. The length of the ski is important. In general, when the soldier stands upright and stretches his arm above his head, the tip of the ski should reach to the middle of his palm. After the bindings have been perfectly adjusted to the boots, and the skis have been waxed, he is ready to set out.

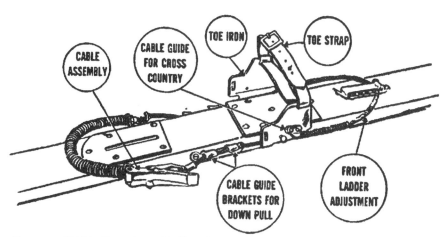

Figure 28. Ski bindings. Some bindings have a front throw to tighten the heel spring.

(1) *Bindings, Ski.* The ski binding (see fig. 28) attaches the foot to the ski and makes it possible to control the ski. The center of the adjustable toe iron of the binding is fastened to the ski at the balance point. The front ladder adjustment should be screwed to the top of the ski far enough ahead of the toe iron so that the cable will be tight around the heel. The cable guide brackets are attached to the side of the ski immediately behind the toe irons. Foot plates protect the surface of the ski under the boot. The only part of the binding which is not permanently fixed to the ski is the cable assembly.

(2) *Adjustment of toe iron.* Ski-mountain boots are the type always worn with skis, since other types of boots do not permit adequate control. The adjustment of the toe irons is extremely important. Each should be fitted to the boot so that the boot rests absolutely straight on

the ski in the exact center. The boot should not extend far in front of the toe iron; about 1 inch is the maximum permissible. To adjust the toe irons, all the screws are loosened and the five center ones removed. After the boot has been placed lightly in position, the cover plate is raised just enough to disengage the adjustment teeth. As the side plates of the toe iron then slide easily, they may be pushed to fit the sides of the boot soles exactly. After the cover plate is pushed down to engage the teeth again, the boot is removed carefully so that the adjustment will not be affected. The screws on the front and rear bands of the cover plate are tightened, then the other screws are replaced and also screwed tight. The toe strap is then inserted, the loose end extending toward the outside to indicate whether it is the right or left ski.

(3) *Adjustment of cable assembly.* The length of the cable is determined by the front ladder adjustment. Under the cover, which is hinged in front and may be opened by lifting it from the rear, there are four steps over which the cable may run. If the cable slips onto the heel too loosely, it should be placed farther ahead. If it is too tight, it should be run over one of the ladder steps farther to the rear. Only when these means of fitting prove inadequate, should the whole front ladder adjustment be moved ahead or farther back. When the cable is in place, the cover is snapped down. There are two possible attachments for the cable next to the foot. For climbing up hill or for cross-country skiing, the cables are placed in the attachments on the side of the toe iron, which allows the heel to move easily off the footplate. For down hill skiing, however, control over the skis is better if there is sufficient down pull to prevent the heels from rising easily. Therefore when skiing down hill, the cables are loosened, slipped under one or both hooks of the cable guide brackets on the side of the ski, and then retightened. If both hooks are used, there will be greater down pull.

(4) *Attaching ski to foot.* The ski is easily attached to the foot. The boot is slipped into the binding to which it has been adjusted, the spring of the cable is placed in the groove on the heel of the boot and the cable is tightened. Some cables are tightened by means of a throw on the front ladder attachment. Others have a tightening lever on the side of the cable, just ahead of the heel spring. The cable should be placed in the front ladder attachment and in the cable brackets in such a way that this lever rests on the outer side of the heel. To utilize fully the safety feature of the combination heel spring and tightening lever, it is important that the cable be placed in the front ladder in such a way that the lever snaps open automatically on a forward fall. If the lever is placed too far behind the heel, it locks and will not open; if it is placed too far forward, it will not stay closed. Each man should find the place, between these extremes, at which the lever will snap open when the leg is bent too far forward.

(5) *Strap, Ski, Safety.* Ski safety straps are provided to keep skis from running down the slope if they come off in a bad fall. The slit end of the strap is run through the loop above the heel of the boot. The buckle is threaded through the slit and then hooked onto the clevis on the inner side of the cable just ahead of the heel spring.

b. Care of skis. (1) *During a halt,* the skis are usually removed from the feet and stuck into the snow, heels down, unless upright skis would give the position of the group away to the enemy. Sometimes the skis are used to sit on when the skier wants to keep dry. To wax skis in fine weather, any snow should be wiped off, and they should be left, running surfaces exposed to the sun, with their heels on the snow and their tips held in the wrist loops of upright ski-poles.

(2) *At the end of the day,* skis are cleaned of snow, strapped together, and placed heels down in their assigned places, preferably under cover. If left outdoors, they should be set with their heels on the dry ground or hard snow where no dripping water will strike them. In warm weather, they are exposed to the sun for a few minutes to dry them and then placed in the shade. Skis must never be left on their sides or they will warp.

(3) *When camping on snow,* skis should be left upright to keep them from getting buried in a snowstorm or drifted over. When stuck into wet snow, they must be absolutely straight up and down. Otherwise their own weight will warp them. They may be laid against trees in the woods.

(4) *Maintaining correct bend.* The skis leave the factory with exactly the correct bend and it is therefore a mistake to tie them together under tension, unless they begin to flatten out. If the skis start to lose their camber or arch, they should be strapped together at the heels and tips with a block of wood about $1\frac{1}{2}$ to 2 inches thick inserted under the binding. The strap near the tip is placed at the beginning of the upturn, so as not to change the bend of the tip in the least. If the tip begins to lose its upturn, the skis should be strapped together as described above. A cord should then be slipped through the hole in the tip, pulled back until the original bend is regained, and tied to the front ladder adjustment of the binding.

(5) *Maintaining and restoring finish.* If any part of the wood of the ski gets exposed from lack of wax or finish, it is likely to warp, because the exposed part soaks up water faster while the ski is in use and dries out more quickly when it is not. This warping and twisting is caused by the uneven increase or loss of moisture throughout the ski. The top edges of the ski become worn from frequent scraping by the steel edges of the other ski until the finish is off. In wet snow, this exposed wood will soak up moisture which will make it wear even faster. The tips of the steel edges may then get uncovered along the top edges of the ski which will allow them to pull out. It is extremely important,

therefore, to keep the finish of the ski in good repair. In addition, too much moisture or exposure to the sun, as well as sudden changes in the temperature, will tend to crack or check the finish on the tops of the skis.

(a) *Whether or not the skis are in use*, keep the running surfaces well waxed. The wax wears particularly quick near the edge of the front quarter of the ski and on the heel. Ski wax is necessary not only to help sliding and climbing, but also to keep the skis waterproof and help retain their shape.

(b) *If there are occasional scratches*, especially if they are on the top edges, they should be lacquered or, if lacquer is not immediately available, covered temporarily with ski wax.

(c) *If the whole finish is in bad condition*, the skis should be scraped down and sanded with a medium fine sandpaper, covered with white paint, dried, and then covered with a waterproof lacquer. After drying, the new finish should be rubbed lightly with steel wool to dull it.

(d) *When skis are to be stored for a long period*, the remaining wax is scraped off and two or three coats of boiled linseed oil, cut with turpentine, are applied to the running surfaces, over a period of about a week. The skis are then strapped with their running surfaces together, and a block about $1\frac{1}{2}$ inches thick between the skis at the bindings to maintain the camber. They are set on their heels in a cool, dry place and must stand absolutely perpendicular, not at an angle.

c. **Repairs.** (1) *Kit, Repair, Ski.* Ski repair kits are issued for minor repairs.

(a) The four-man kit contains the following items:
- 1 combination file and countersink.
- 1 scraper.
- 2 pairs of combination pliers, with wire-cutters and milled laws, a screw driver on one handle and a wrench on the other.
- 12 No. 2 wood screws with No. 2 recessed heads, $\frac{1}{4}$ inch long.
- 12 No. 3 wood screws with No. 3 recessed heads, $\frac{1}{4}$ inch long.
- 12 No. 2 wood screws with No. 2 recessed heads, $\frac{5}{16}$ inch long.
- 12 No. 3 wood screws with No. 3 recessed heads, $\frac{5}{16}$ inch long.
- 12 No. 3 wire size bolts with hexagonal nuts, $\frac{1}{2}$ inch long.
- 6 6-inch sections of steel edges.
- 10 feet of steel wire.
- 2 $\frac{3}{16}$-inch leather laces 55 inches long.
- 2 $\frac{1}{2}$-inch leather laces 55 inches long (for emergency binding heel strap).

(b) As soon as any items have been used, they should be replaced. The mountain pocketknife is fitted with other tools needed for emergency repairs. No wood drill is provided. If a small hole is needed while in the field, a nail or bolt can be heated and allowed to burn its way through the wood.

(2) *Repairs to steel edges.* (a) *Edge screws* should be tightened if they work loose, but care is necessary to keep from breaking their shanks off and from twisting the threads out of the wood. If a screw breaks off, another should be inserted at a slight angle and its head filed flat. If the threads in the wood are pulled out, a larger (No. 3) screw is used. To make the head of the screw flush with the surface of the steel edge, the hole in edge is reamed out with the countersink on the end of the file before the screw is inserted. If the hole is clogged with broken screws, or is too large for the biggest screws, it must be drilled through for a No. 3 wire size ½-inch bolt. After the nut is screwed down tightly on the upper surface of the ski, the part of the bolt extending beyond the nut is filed off. Skis should be inspected daily for lost edge screws.

(b) *If the steel edges are humped up between the screws,* they should be removed and straightened on a wooden block with light hammer blows, until they are bent back to their original shape. The wood on which the edges rest should be cleaned of any foreign matter before they are screwed back on.

(c) *If a section of the steel edge is missing,* it should be replaced at the first opportunity. The whole broken section should be removed. If one or two of the sections in the ski repair kit fit the gap perfectly, the new section of edge is then screwed on. Sometimes, however, part of a section must be fitted in. Sections are cut off just beyond the screw holes so that they will fit exactly. Where two cut sections join, they should be beveled so that the forward part overlaps the rear. The whole edge is then screwed down.

(d) For making emergency repairs to *broken skis,* see paragraph 79.

(3) *Repairs to ski bindings and replacements.* (a) *All screws* should be inspected and tightened from time to time. If a particular screw loosens often, it should be replaced by a slightly larger one.

(b) *Toe straps,* which wear particularly in the slot in the toe iron and at the adjustment holes, should be replaced before they break.

(c) *Hinge pins* eventually wear where the cables are tightened to a point where they need to be replaced.

(d) *The cables* should be replaced when they show excessive wear at the front ladder adjustment, at the cable brackets, or in front of the heel spring, as well as when the housing is broken or worn through. Cables occasionally break in bad falls. The ½-inch strap from the ski repair kit is then used as an emergency heel strap until the skier returns to his base. To improvise an emergency heel strap, cut a small slit in one end of this ½-inch strap and run it through the vertical slot on the rear of one side of the toe iron, around the heel groove, through the slot on the other side of the toe iron, and back around the heel. One end is then run through the slit in the other end. The straps are pulled up as tightly as possible and fastened with two half-hitches on a sheet bend. The loose end of the piece of lace may be placed over the instep on both sides and pulled up to take in any slack.

76. POLES, SKI, STEEL TYPE II, LAMINATED TYPE I. a. Nature and use. Both laminated cane and steel ski-poles are issued to troops operating on skis. They are used primarily for poling to help the skier move faster and more efficiently along the level and uphill, but they serve other useful purposes too. They can be used, for example, as tent poles, as rafters for snow shelters (improvised), as gee-poles (when lashed onto the front of a sledge), or, with rings removed, as a substitute for avalanche probes.

   b. **Fitting.** There are four ski-pole sizes. A ski-pole should ordinarily reach about up to the skier's armpits when set on the ground next to him.

   c. **Care and repair.** (1) *Cane poles.* Cane poles must never be brought directly out of a hot room, for sudden changes of temperature may cause them to split.

   (2) *Wrist straps.* Wrist straps break occasionally in bad falls. The rivet which holds the strap to the top of the pole, because it is covered by the grip, is difficult to get at if the leather pulls away from it. In such a case, holes should be punched in the attached portion of the leather near the grip and on the end of the strap. A piece of rawhide or string is then threaded through the holes and tied to hold the new loop together. When the soldier returns to his base, the grip is unthreaded, the strap reriveted, and the grip resewn.

   (3) *Rings.* A ring which is damaged, badly worn, or lost should be replaced immediately.

77. WAX, SKI. a. Importance. Ski wax serves two major functions:

   (1) *Aid to operation.* Ski wax is applied to the running surfaces of skis to make them slide down hill and to keep them from slipping back when climbing up hill. The same coating serves both purposes if the skis are properly waxed. Climbing roughens the wax. If the surface of the snow were magnified, tiny points could be seen on the snow crystals. When the ski is pressed down on these points, they bite into the surface of the wax and keep the skis from slipping back. When the ski slides forward the wax is smoothed out and the points of the crystals have no chance to hold the ski back. For long climbs, it is more efficient and less tiring to use ski climbers than to use wax alone, but when the going is both up and down hill, or when there is risk that the enemy may not allow a unit time enough to put climbers on, it is best to wax for up hill as well as down hill skiing.

   (2) The importance of waterproofing skis to avoid warping is discussed in paragraph 75b(5). Skis must be waxed, whether or not they are in use.

   b. **Waxing.** It is easiest to wax skis in a warm place, for the wax will spread most easily and evenly if it and the skis are moderately warm. New skis have treated running surfaces on which the wax is applied.

The first time the skis are waxed, a layer of wax should be rubbed over the whole surface and smoothed down. The skis should then be put in a cool place to allow the layer to harden before the surface wax is applied. If the skis have already been used, the old wax is rubbed down smooth and used as a base wax. Although the wax must never be allowed to get too thick, the old wax has to be removed only in rare instances. It takes experience to become an expert on wax. Because of the large number of variables, only a few hints on the specific use of waxes can be given here:

(1) *For dry snow* (snow which will not make snowballs). Polish a thin layer of blue dry snow wax on the skis until smooth. If skis are slow, add a thin layer of red speed wax and polish it for down hill running. If the skis will not climb, apply a rougher layer of blue. If they still do not climb well, apply a thin section, about 18 inches long, of rough orange wet and corn snow wax under the feet.

(2) *For wet snow* (snow which makes excellent snowballs). Apply a thin layer of orange and smooth it out only slightly. If this does not make skis run well down hill, use red speed wax fairly rough, or paraffin, if available. If the skis do not climb, apply an 18-inch section of rough orange under the feet.

(3) *For corn snow* (granular snow). Wax about the same as for wet snow, but apply the wax in slightly thicker layers.

78. CLIMBERS, SKI, MOHAIR. a. Purpose and nature. Ski climbers are used on the running surfaces of skis to facilitate climbing. The mohair lies with the ends pointing toward the heels of the skis. When the skis slide forward, the hair remains flat, offering little friction; but when the skis start to slide back, the hair is rubbed the wrong way, becomes roughened, and keeps the skis from slipping. Ski climbers are used when the ascent is sufficiently long to justify taking the time to put them on.

b. Manner of use. The big loop near the rear end of the climber is slipped over the heel of the ski so that the straps and the buckle rest on its upper surface. The front loop is placed over the ski tip and the strap on the rear end of the climber is run over the ski heel and back forward along the top of the ski to the buckle in the big rear loop. It is pulled as tightly as possible, and then fastened. The other two straps, one just ahead of the bindings and the other just behind them, are pulled up tight and buckled. Straps must be kept tight to keep the climbers from slipping off. Some ski climbers are fitted with three bar buckles with V-shaped sides. These may be used in extremely cold weather without removing the mittens. The strap is run under the two exposed bars, back over the middle one at the point of the V, and under the end one. The strap is tightened by pulling the end of it. The buckle can be easily released by pushing the point of the V toward the rivet.

c. **Care.** Climbers will last longer if properly cared for. The following are important precautions:

(1) Climbers should be removed before a descent of any length, as down hill skiing seriously shortens their lives.

(2) Straps must be kept in good repair and stitches reinforced whenever necessary.

(3) When not in use, ski climbers are usually rolled up; but if they are wet, they must be unrolled and hung up to dry at the first opportunity.

(4) When the rucksack is not carried, the climbers are worn around the waist like a belt.

79. TIP, SKI, EMERGENCY REPAIR. a. Purpose. An emergency repair ski tip makes it possible to get back to base after snapping off the tip of a ski. Since such a repair is makeshift, an effort must be made at the first opportunity to repair the ski properly or to replace it entirely.

b. **Types.** There are two types of emergency repair ski tips: wooden and metal.

(1) *Use of wooden emergency repair ski tip.* If the ski is badly shattered, the slivered end should be trimmed off fairly even. The wooden emergency repair ski tip is set under the broken end with sufficient overlap to bring all the screwholes over solid wood. Four steel edge screws, either removed from the broken tip or taken from the ski repair kit, are screwed up into the lower surface of the ski though the screwholes in the emergency repair ski tip. To reinforce the joint further, two contraction bands are then placed over the overlap.

(2) *Use of metal emergency repair ski tip*.* This type of tip may be used in two ways:

(a) The most successful repairs are made by holding the broken ski tip in place with the emergency repair tip or the contraction band. When the broken tip is used, the ends are fitted back together. Then the emergency repair tip is placed under the ski with the channels overlapping the break, so that the forward screwholes are ahead and the rear ones behind. Four steel edge screws, either removed from the broken tip or taken from the ski repair kit, hold the tip firmly in place.

(b) If it is lost or badly shattered, the broken tip obviously cannot be used for repairs; nor can it be used unless the break is about 8 to 12 inches from the point of the ski. In such cases, the emergency repair tip is used alone. The tip is slipped onto the broken remainder of the ski so that the break rests just ahead of the front of the channel and is held firmly in place by four steel edge screws. In this case, as above, the width of the emergency repair tip, which is adjustable, should be kept to a minimum so that the ski will be firmly seated in the channel.

80. BAND, CONTRACTION. The contraction band is useful in making emergency repairs if a ski tip is snapped off. The broken surface of the tip is trimmed down to allow the two surfaces to meet smoothly without a definite step when the tip is placed under the main part of the ski. The tip should overlap the rest of the ski by at least 3 inches more if it is badly splintered. Two contraction bands are placed around the overlapping section, and after the contraction bolts have been threaded into the two toggle nuts, they are screwed tightly together. The forward edge of the first contraction band should rest on a smooth surface on the underside of the tip, so that snow will not force its way underneath the band. If possible, cover the bands with tape, and wax the part which is on the sliding-surface of the ski. If it is impossible to use two contraction bands, one, carefully adjusted, may be used alone.

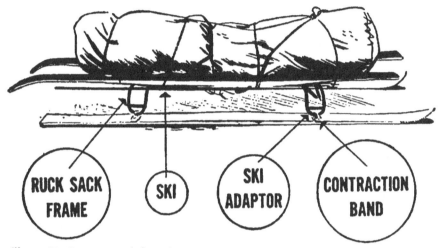

Figure 29. Emergency sled made from four pairs of ski adaptors, four contraction bands, two pairs of skis, and two rucksack frames.

81. ADAPTOR, SKI. *a.* **Purpose.** Ski adaptors make it possible to improvise an emergency sled quickly from equipment usually carried by the skier. The necessary equipment includes—

    4 pairs of ski adaptors.
    4 contraction bands.
    2 pairs of skis.
    2 rucksack frames.

b. **Setting up emergency sled** (fig. 29). (1) Place two skis on the snow parallel to each other, a rucksack's width apart.

(2) Open four contraction bands so that the contraction bolts remain screwed into the toggle nuts on only one side. Slip the contraction bands under the skis, one about 18 inches from the tip and one the same distance from the heel.

(3) Remove two rucksack frames from the rucksacks. Place these on the skis so that the parts of the frame which normally extend forward

near the hips rest on the upper surface of the ski, immediately above the contraction bands. The parts of the frame which are normally at the base of the neck will nearly meet above the ski bindings.

(4) Place the ski adaptors on the skis with the flat surface down, one piece on each side of the metal tubing of the rucksack frame and properly aligned, so that the channels will fit the tubing. Push the ski adaptors together so that they grip the rucksack frame tubing firmly.

(5) Thread the contraction bolts across the ski adaptor and the rucksack frames into the toggle nuts on the opposite side. Screw the bolts tight.

(6) Lash together the rucksack frames where they meet above the bindings, and reinforce the joint by means of ski-poles or sticks placed diagonally through the top bends of the rucksack frames.

(7) Lash a pair of skis onto the rucksack frames to make the platform for the sled. Ski-poles, sticks, or another ski will improve the platform.

c. **Precautionary measure with injured.** When the sled is used for an evacuation, the casualty must be lashed on with particular care. Ski climbers are well adapted to the purpose of lashing the patient, since they are broad, and will not cut into him and restrict his circulation. The patient should be kept as warm as possible, preferably in a sleeping bag, to reduce the chances of frostbite and shock. More padding should be placed beneath than on top of him, as a protection both against the cold coming up from below and from any roughness on the platform.

d. **Trace ropes.** Traces improvised from tent ropes, climbing ropes, etc., are attached to the holes in the ski tips. For level going and up hill climbing, ropes are needed only in front of the sled; but for steep descents, other ropes must be attached to hold the sled back.

e. **Gee-pole for rough going.** A gee-pole, which is extremely useful to guide the sled in rough country, is easily improvised. One end of a stick or pole is lashed firmly at the right of the front rucksack frame where it rests on the ski, and the pole is then lashed again to the right side of the platform so that it is held forward at a gentle upward angle. When the man who is to pull next to the sled takes his place in the traces, the gee-pole will be beside his right hand where he can easily grasp it to guide and steady the sled or to turn it quickly. An additional pole on the left side is useful for guiding the sled down gentle ski slopes. The skier steps between the two poles, grasping one in each hand, so that he can pull the sled along, hold it back, or steer it.

82. SKIS, CROSS-COUNTRY. Cross-country skis are issued for use in flat or rolling terrain. They have no steel edges. These skis are lighter and slightly narrower than steel edged skis. They come only in the 7-foot 3-inch length. Flexible ski bindings should be used with them.

83. BINDING, SKI, FLEXIBLE. a. General. Flexible ski bindings are used on cross-country skis with any type of winter footgear such as shoepacs, mukluks, felt shoes, etc.

b. Mountain binding. (1) Place the boot along the balata strip with the heel in the heel cup. Center the strip and boot on the ski with the ball of the foot over the balance point of the ski. (This is marked on the side of the ski.) Remove the boot, holding the strip firm.

(2) Place the metal cover plate on the balata strip so that the two sets of holes in the cover plate lie on either side of the balance point. Mark the holes through the plate and strip. Remove and check marks to see that they are equally distant from either side. Drill holes $\frac{3}{32}$ inch in diameter and $\frac{1}{2}$ inch deep.

(3) Insert toe piece wiith straight edge toward the ski tip and place the cover plate over the strip. Match the holes in the ski, toe piece, strips, and cover plate. Screw down all four screws.

(4) Drill holes in the ski to match the front two holes in the balata strip. Screw down.

c. Use. To adjust the binding, buckle the toe piece over the toe of the boot and the ankle strap over the instep. If the boot is properly fitted, the toe of the boot is held snugly by the toe piece and the heel rests flush in the heel cup. If snow sticks between the ski and the balata strip, rub paraffin or red ski wax on the strip.

Section VII. SNOWSHOES

84. PURPOSE. Snowshoes as well as skis are used for operations on snow. Skis are usually faster and less tiring, particularly for downhill running, but snowshoes require less skill to handle, and are easier to manage in rough or thickly wooded terrain. Snowshoes are also more efficient when carrying heavy loads and when pulling sleds. In deep snow they are frequently used to break trail for sledge dogs.

85. USE. a. Technique. It takes only a little practice to become skillful in the use of snowshoes. The beginner soon learns how to avoid catching the tip in the snow and tripping himself, and acquires the waddling gait needed to keep from stepping on the side of one snowshoe with the side of the other. On a steep descent he can slide by squatting down, putting the toe of one snowshoe on the other and holding it under the heel of the other boot.

b. Pad, Snowshoe, Mountain. Although soft-soled boots are the type ordinarily worn with snowshoes, nailed ski mountain boots can be used with *mountain snowshoe pads*. If pads are not used, the nails will wear right through the webbing. The pad is lashed onto the snowshoe under the foot. There are six sets of three holes each, located at the corners and on both sides of the pads. Two of the holes in each set should lie above some strand of the webbing where they will be convenient for holding the lashing. The pad is revarnished when the need arises.

86. CARE. Snowshoes must always be kept in good condition. Whenever a piece of rawhide breaks, it should be repaired as soon as possible or the whole webbing will loosen and unravel. Sometimes in wet snow, the rawhide begins to get white, particularly at the edges and at other points of wear. At the first opportunity, the snowshoes should be dried slowly for a period of about 48 hours in a cool, dry place where air is circulating. Then the whole of the snowshoe, both webbing and frame, should be given one or two coats of waterproof varnish. At the end of the winter or halfway through a long season, even though the need is not yet apparent, snowshoes should be revarnished. When the wooden frames fray, the splinters should be carefully removed, the edges smoothed, and the whole frame revarnished. All leather bindings should be treated with dubbin once or twice a winter. Snowshoes must be stored in a dry place so the rawhide will not mildew or rot, and the frames will not warp. Even if they are to be left for only a day or two, they must be hung up where rodents cannot gnaw the rawhide webbing. This is equally important in the field.

Figure 30. Trail snowshoes. (Note binding detail.)

87. SNOWSHOES, TRAIL. Since trail snowshoes are long and narrow with upturned toes, they are particularly good for fast cross-country travel in open and sparsely wooded terrain. The binding permits the heel of the boot to rise off the snowshoe, but allows the tail to drag on the ground like a toe slipper. The ball of the foot rests on the webbing just behind the toe hole, so that the front of the boot pushes through when the wearer lifts his heels or when he stands on his toes to get better traction. The binding is attached to the snowshoe (see fig. 30) in the following manner (the outer side is always the left of the left snowshoe and the right of the right snowshoe):

a. Lay the toe cap (the short wide strap with four slits) along the center of the reinforced cross webbing, just behind the toe hole.

b. Thread the toe strap (the longer strap with the buckle) down through the outer rear slit around the reinforced cross-webbing, back up through the outer front slit, across the toe cap, down through the inner front slit, around the reinforced cross-webbing, and back up through the inner rear slit. Buckle the strap together.

c. Thread the heel strap which has a buckle (the shorter of the two straps with buckles) down in front of the reinforced cross-webbing (between the two outer lengthwise strands of webbing beside the toe hole) and back around the cross-webbing. Open the slit and feed the buckle through.

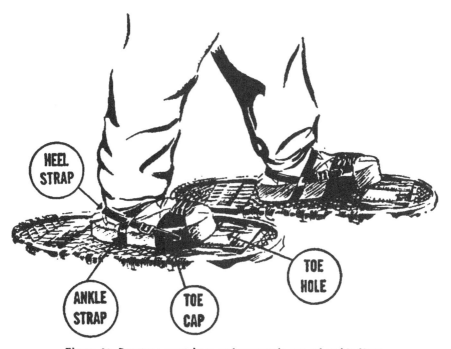

Figure 31. Bearpaw snowshoes and mountain snowshoe bindings.
(Note binding details.)

d. The heel strap which does not have a buckle (the longer strap without a buckle) is threaded similarly, but on the inner side. Then the slit is opened and the end with the adjustment holes is fed through it.

e. Slip the long heel strap through the loops on the ankle strap so that the adjustment holes and buckles will meet above the instep. Once the binding is on the snowshoes, it does not have to be removed again. The foot is slipped under the toe cap until the ball of the foot rests on the reinforced cross-webbing. The top strap is then adjusted to hold the foot firmly, but not tight enough to cut off the circulation. The heel strap is buckled around the ankle, high enough so that it will stay on, but not high enough to hurt the Achilles tendon. The ankle strap goes completely around the foot and is fastened loosely above the instep to hold the heel strap in place. Just how tight the straps should be can be determined from experience. If they are too tight, the circulation of the blood will be restricted so that the feet will become cold or frostbitten. If they are too loose, the snowshoes will fall off the boots. Worn binding parts should be replaced as soon as possible.

88. SNOWSHOES, BEARPAW. For hard going in thick woods and rough, difficult terrain, Bearpaw snowshoes (see fig. 31) are handier than the longer, narrower trail snowshoes. They have an oval shape and no tails. Bearpaw snowshoes are held onto the boots by mountain snowshoe bindings. They are lashed firmly onto the center of the reinforced cross-webbing, just behind the toe hole. The foot is slipped into the toe cap so that the ball of the foot rests on this cross-webbing, and the lacings are tightened to hold the foot from sliding farther through. The heel strap is buckled around the foot, high enough to keep it from slipping off but not high enough to press on the tendon. The ankle strap holds the heel strap in place. In a properly adjusted binding, when the heel is raised off the snowshoe, the toe extends through the toe hole and gives good traction for hauling loads or climbing steep slopes. The claw will hold on crust.

Section VIII. MISCELLANEOUS EQUIPMENT

89. BOX, MATCH, WATERPROOF. a. Description and purpose. The waterproof matchbox is a small, cylindrical, plastic box with a screw cap. Mobile troops operating away from their bases need dry matches for starting fires.

b. Use. The matchbox holds about 20 ordinary "strike anywhere" matches and a larger number of safety matches. Double the number of matches may be carried, if they are cut in half and only the tip end is used. Ordinarily, matches carried in the box are reserved for lighting fires, etc. Matches for smoking are carried separately to keep the supply of matches in the matchbox for real emergencies.

c. Striking matches. "Strike anywhere" matches may be struck on a dry surface or along the special grooves on the side of the matchbox. There is a small striking disk inside the cap for safety matches.

d. Striking bar. To save matches, the metal bar on the bottom may be struck with a knife blade, screw driver, or other metal object. The resulting spark is enough to start a gasoline stove or a heap of tinder.

Tinder usually consists of fine dry grass, leaves, splinters, lint, or cotton. The powder from a cartridge may also be used. A bit of gasoline poured onto the tinder will make it light more easily. When the spark has set the tinder on fire, blow on it until it bursts into a good flame. An adequate supply of fuel must be on hand to keep the fire going.

90. CREEPERS, ICE. a. Purpose and nature. Ice creepers may be worn with any type of shoe except nailed or rubber-cleated ski-mountain boots, which grip adequately on average ice and snow. The metal spikes bite into hard snow and ice enabling troops to move safely and efficiently on such terrain. They are *not* a substitute for crampons, which are used for steep climbing on hard snow and ice.

b. Use. The ice creeper is strapped on the shoe. It is placed on the foot, spikes down, with the chain toward the heel. When the ice creeper has been adjusted to fit the front of the shoe, the D-rings which hold the straps are bent over the edge of the sole and the straps buckled across the toe. The straps in the rear, which are attached to the end of the chain, run around the back of the heel and are buckled over the instep.

91. BRUSH, MOUNTAIN. The mountain brush is particularly useful for troops camped on snow. Before entering a tent, each soldier brushes off any snow clinging to his trousers, socks, boots, rucksack, and other objects. As soon as he has removed his outer clothing, socks, and boots, he brushes out all the frost he can. He naturally sweeps snow and loose frost out the tent door before they have any chance to melt and wet the sleeping bags and clothing. In terrain where dirt is tracked into the tent, the mountain brush is used to keep the floor clean.

92. MARKERS, TRAIL. Trail markers are used for marking trails on snow. Although they must not be used where they will give positions or routes away to the enemy, they are essential for indicating difficult routes, especially in bad weather and at night. For example, they may be used to show the best line through the crevasses on a glacier. Trail markers are ordinarily placed about 100 feet apart, or at bends in the routes. In bad weather, one man should stay at a marker until the man ahead finds the next one.

93. CREAM, SUNBURN-PREVENTIVE. a. Need. On snow, particularly at high altitudes, sunburn is often serious enough to keep men from performing their usual duties. Minor cases may be extremely painful, and bad ones are accompanied by running sores and infections. In the snow, sun not only shines directly on the exposed skin, it is also reflected up from the snow below. Since the diffused rays of the sun burn just as badly in bright foggy weather, precautions must be taken even when the sun is not shining.

b. Preventive measures. The best way to avoid sunburn is to keep covered. Shirts and hats should not be removed in brilliant sunshine. Exposed portions of the body, such as the face, must be protected by rubbing sunburn preventive, issued to troops who are in danger of being badly sunburned, into the skin before exposure. It should be renewed during the course of the day.

INDEX

	Paragraph	Page
Adaptor, ski	81	73
Ax, ice, mountain	70	58
Bag, food, waterproof	65	54
Bag, sleeping: wool; mountain; Arctic; kapok	58	43
Band, contraction	80	72
Boots	30	20
Blucher, high top	32	24
Ski-mountain	31	23
Box, match, waterproof	89	78
Breaking camp	69	57
Brush, mountain	91	79
Cap:		
Field, cotton, OD	46	31
Field, pile, OD	45	31
Ski	47	32
Wool, knit, M1941	49	32
Cleanliness	6	2
Climbers, ski mohair	78	71
Cold-weather handwear	38	28
Container, fuel, 1-quart	64	54
Cook set, mountain	63	54
Crampons, mountain	71	60
Cream, sunburn-preventive	93	79
Creepers, ice	90	79
Drying clothing in cold weather	9	3
Essential characteristics of cold-weather clothing	8	2
Gaiter, ski	37	28
Gloves:		
Insert, wool	42	30
Shell, leather	41	30
Goggles, ski-mountain	53	34
Helmets:		
Combat, winter	50	32
Steel	51	33
Hood, jacket, field M1943	54	34
Insoles	29	19
Jackets:		
Combat, winter	20	16
Field, M1943	11	5
Field, OD, Arctic	21	16
Field, pile	12	8
Mountain	18	14
Markers, trail	92	79
Mask, field, chamois	52	33

	Paragraph	Page
Mittens:		
Insert, trigger finger	40	30
Shell, trigger finger	39	29
Mountain cook set	63	54
Mountain crampons	71	60
Overcoat, parka type, with pile liner	17	14
Overshoes, Arctic	34	26
Packboard	57	41
Pack, field	55	35
Pad, insulating, sleeping	59	47
Parka:		
Field, cotton, OD	13	8
Field, over, white	15.	12
Field, pile	14	11
Reversible, ski	19	15
Wet weather	16	12
Personal precautions	68	57
Perspiration, importance of avoiding	7	2
Pitching tent	67	56
Pitons	73	62
Poles, ski	76	70
Reference	3	1
Ropes, climbing	72	61
Rucksack	56	38
Shoepacs	33	25
Shoes, Arctic, felt	35	26
Shoes, felt, cold-weather	36	27
Skis	75	65
Adaptor	81	73
Gaiter	37	28
Poles	76	70
Tip, ski, emergency repair	79	72
Wax	77	70
Socks	28	18
Stove, gasoline, one-burner, M1941 and M1942	62	49
Tent, mountain, two-man complete	66	55
Toque, wool knit, M1941	48	32
Trousers:		
Combat, winter	26	17
Field, cotton, OD	22	16
Field, over, white	24	17
Field, wool, OD	23	17
Kersey-lined	26	17
Mountain	25	17
Wet weather	16	12
Wax, ski	77	70

Made in the USA
Coppell, TX
22 November 2023